Giacomo Grazzi-Soncini

# Wine

Classification, Tasting, Qualities and Defects

Giacomo Grazzi-Soncini

**Wine**
*Classification, Tasting, Qualities and Defects*

ISBN/EAN: 9783743372962

Manufactured in Europe, USA, Canada, Australia, Japa

Cover: Foto ©berggeist007 / pixelio.de

Manufactured and distributed by brebook publishing software
(www.brebook.com)

Giacomo Grazzi-Soncini

**Wine**

APPENDIX E TO THE BIENNIAL REPORT OF THE BOARD OF STATE
VITICULTURAL COMMISSIONERS FOR 1891-92.

# WINE.

## CLASSIFICATION—WINE TASTING—QUALITIES AND DEFECTS.

By

### PROF. G. GRAZZI-SONCINI,
Director of the Royal School of Viticulture, Alba, Italy.

TRANSLATED BY

### F. T. BIOLETTI,
Of the Agricultural Experiment Station (Viticultural Section), University of California,
Berkeley, California.

SACRAMENTO:
STATE OFFICE, : : : : : A. J. JOHNSTON, SUPT. STATE PRINTING.
1892.

# OFFICERS AND MEMBERS OF THE BOARD.

GEORGE WEST, President.......................................................Stockton.
Commissioner for the San Joaquin District.

CHARLES BUNDSCHU, Vice-President ...........................San Francisco.
Commissioner for the San Francisco District.

ALLEN TOWLE, Treasurer ......................................................Towles.
Commissioner for the El Dorado District.

J. DeBARTH SHORB.....................................................San Gabriel.
Commissioner for the State at Large.

JOHN T. DOYLE ....................................................San Francisco.
Commissioner for the State at Large.

ISAAC DeTURK..............................................................Santa Rosa.
Commissioner for the Sonoma District.

E. C. PRIBER ....................................................................Napa.
Commissioner for the Napa District.

R. D. STEPHENS .........................................................Sacramento.
Commissioner for the Sacramento District.

E. C. BICHOWSKY.....................................................San Gabriel.
Commissioner for the Los Angeles District.

WINFIELD SCOTT, Secretary................................San Francisco.
CLARENCE J. WETMORE, Chief Executive Viticultural and Health Officer....
......................................................Livermore and San Francisco.

*Office of the Board:*
317 PINE STREET, SAN FRANCISCO.

# AUTHOR'S PREFACE.

Royal School of Viticulture and Œnology,
Alba, Piedmont, Italy, January, 1892.

A preface should give an immediate idea of what the author has proposed to do in writing his book. As Balbo rightly says in the preface to one of his books:

"It is the duty of every writer to give the reader a terse and clear idea of the work which he presents him. This sincerity benefits both: the reader, because it puts him in the position of knowing whether or not the book is likely to be of interest or utility to him; the writer, because, whilst it may reduce the number of his readers, it insures him more interested, attentive, and indulgent ones.

"The clearest and most sincere way of giving an explication of the object of a book is to tell how it was written."

Thus I will explain, as well as possible in a few words, why I have written this book, which treats especially of the classification, the qualities, and the defects of wine.

When I commenced to give particular attention to viticulture and œnology, I soon perceived that in œnology, and especially in that part which regards classification, qualities, and defects, all authors were not in accord in their use of terms to express the same characters. Thus, for example, some would mean by "sève," a slight sweetness in the wine; others by the same term would intend to express that character by which a wine of good quality affects the mouth and olfactory organs with a certain perfume, for a longer or shorter time after it has been swallowed.

I will say nothing of the classification of wines according to dishes, as wine to be drunk with oysters, fish, roast meat, etc., which shows a marked tendency to become a veritable chaos. In this classification, the work of Mr. Bertall, "La Vigne-Voyage Autour des Vins de France," is taken too literally.

How could one speak of the classification of wine, of its qualities, of its defects, without giving some explanation of the mode and proper conditions for tasting? It is for this reason that I have devoted a chapter to the tasting of wine, a chapter, moreover, of great importance, as it is by tasting, more than by chemical analysis, that we can best judge of the constitution and future of a wine. Who is a better judge than an experienced taster of the bad flavor produced in wine, for instance, by the tartaric fermentation, which even in its incipiency he can detect by a certain burnt taste, which, with the progress of the malady, gradually develops into an insupportable bitterness? Among these gradations of bitterness we do not find that slight pleasing bitterness peculiar to certain wines, such as Barolo and Gattinara.

Chemical analysis gives us the principal components of wine, and from the presence or absence of certain of these and from their proportions, some judgment may be formed of the character of the wine. The taster alone is able to detect diseases at their incipiency, and, one might almost say, before they have commenced, whilst the chemist can only state the final consequence. In other words, one might say that whilst the chemist is limited to making a diagnosis, the taster can make a prognosis.

In the case of some defects of wine, I have not confined myself to a simple definition or description. I have also added notes, brief in some cases, more extended in others, on the determining causes and the means of prevention or cure. I have done this, believing it would be useful to the taster or the dealer, who is not always fully informed on all the details of technical œnology. With this information for a guide, he will be better able to judge of the relative gravity of this or that defect, and the dealer especially will be able to judge of the utility or inutility of attempting to cure a wine of a certain defect.

I have also tried—wishing to be useful to the greatest possible number of readers—not to neglect a secondary part, which has its importance in tending to make the consumer better appreciate the wine he drinks. Profiting by the *Consigli di un amatore di vini*, I have indicated the form of glass to be used with each kind of wine, how wines should be presented and distributed during the repast, and how they should be drunk. In this part, which I have called secondary, it is not to be denied that fashion is the determining factor.

And now the reader may judge if I have succeeded in my intentions. Even though his judgment should not be favorable, I shall consider myself fortunate in being the first—as far as I know—to call attention, in an extended manner, to this part of œnology, which in former treatises on the subject has been but lightly touched upon.

G. GRAZZI–SONCINI.

# TRANSLATOR'S PREFACE.

AGRICULTURAL EXPERIMENT STATION, UNIVERSITY OF CALIFORNIA, }
BERKELEY, CAL., September 16, 1892. }

Professor Grazzi-Soncini's book, which has been already translated into French, fills a void in the literature of œnology. The part dealing with the defects of wine, the diseases to which it is subject, and the methods, when such exist, of remedying these diseases, will perhaps be of the most practical value to the wine grower. The part which regards tasting and classification, however, is worthy of careful reading, and many hints may be drawn from it that will be of use towards the attainment of that most desirable object: the production of constant types of wines—an object which is too little studied in California, but on which our hopes of building up a trade in high-class wines very largely depends.

Many of the numerous terms which the French and Italians have invented for the technical consideration of wine it is impossible or difficult to translate into English, and for this reason the translation necessarily lacks some of the scientific precision and clearness of the original. I have however attempted, wherever possible, to give the English equivalent of the term used by the author, and have also given the French term, in this way making a glossary in the three languages, which may possibly be of use in developing a uniform set of technical terms on this subject in our own language.

If this book should be of any use to the wine maker, and especially if it should call the attention of non-wine-drinking people to some of the uses and beauties of wine which they did not suspect, the translator will feel amply repaid for his trouble.

F. T. BIOLETTI.

# WINE AND THE ART OF WINE TASTING.

By G GRAZZI-SONCINI.

## INTRODUCTION.

Wine is simply the juice or must of the grape after it has undergone the process of fermentation.* This may be considered as the most natural and exact definition that can be given of it. It is the definition accepted by the law.

On account of the prevalence of sophistications and the considerable amount of wine that is now made from dried grapes and other saccharine fruits, a more particularized definition of wine is now given; it may be formulated as follows:

By wine is understood that liquid which is obtained by the alcoholic fermentation of the juice or must of fresh grapes. This must may be fermented in contact or not with the pomace or solid portion of the grapes, without, however, the addition of any extraneous substance or even of substances chemically the same as those that the grapes them-

---

*Although as Gautier writes, "Wine is a very complex body, and so delicate that the work of chemists, so far, has been but an outline of what there is to do in the study of it," I think it will be useful, because it will give a more complete idea of the subject of our remarks, to give a list of the principal components of grapes, or must, and of wine:

### A. SOLID BODIES.

*Stems:* Lignose—Tannin—Albuminoids—Organic salts and acids—Mineral salts and acids—Chlorophyll—Gummy matters—Phosphates—Potash, lime, magnesia, silica.

*Skins:* Cellulose—Œnocyanin—Œnorubin—Tannin—Cream of Tartar—Catechin—Quercite (?)—Waxy matters, ferment germs—Etherous and aromatic principles—Nitrogenous substances—Phosphates—Potash, lime, magnesia, iron, silica.

*Pulp:* Cellular parenchyma—Nitrogenous substances—Cream of tartar—Gum, pectin, dextrin (?)—Gases, nitrogen, carbonic acid—Divers salts.

*Seeds:* Lignose—Fatty matters—Nitrogenous substances—Gum—Starch—Phosphates—Divers salts—Tannin.

### B. LIQUID BODIES.

Water—Glucose—Levulose—Divers nitrogenous substances—Saccharose, dulcite—Cream of tartar—Tartrate of calcium—Tartaric, malic, and racemic acids—Halogen acids (traces)—Ammoniacal salts and organic derivatives—Phosphates, sulphates, nitrates—Potash, lime, magnesia.

### C. GASEOUS BODIES.

Carbonic anhydride—Nitrogen—Hydrogen sulphide.

### ELEMENTS OF WINE (RED WINE).

Water—Alcohols: ethylic, propylic, butylic (amylic?), caproic, œnanthilic, caprylic, pelargonic, capric.

Higher alcohols—Glycerine—Isobutyl—Mannite—Glucose—Levulose—Inosin—Gum—Pectic matters—Essential oils—Furfurol—Aldehyde—Acetal.

Ethers: acetic, propionic, butyric, valerianic, caproic, lauric, myristic, palmitic, stearic.

Acids: carbonic, acetic, propionic, butyric, caproic, œnanthylic, caprylic, capric, lauric, myristic, tartaric, racemic, succinic, malic, tannic, sulphuric, nitric, phosphoric, silicic, chlorhydric, fluorhydric. These acids are either free or combined with the bases: potash, soda, lime, magnesia, alumina, iron oxide, manganese, ammonia, volatile bases of the pyridic series.

Albuminoids—Coloring matters.

selves contain. The addition of the latter is considered by many as an adulteration, because it changes the quantitative composition of the must, and consequently of the wine.

Who first made wine is not known. The history of its manufacture, like that of many other fermented beverages, extends back into the mists of ages; nothing, therefore, is known about its first use. Tradition and mythology give several accounts of its first appearance, but they are of a very contradictory nature.

Of one thing we may be sure, and that is that from the first, man has asked himself the question: Is wine a real benefit? A question that has not yet, perhaps, been answered to the satisfaction of some.

Even at the present day it is not possible to give a satisfactory, definite reply to this demand, unless we look at it from an economical standpoint, in which case there can be no doubt of its utility, as it is one of the principal sources of national wealth in every country where the grape can be grown.

We must therefore consider it from this point of view, otherwise its real utility to man might be contested.

It is said that wine incites man to anger, licentiousness, murder, and in general subjects him to a thousand depraving temptations.

"*Il vino e il veleno piu teribile per la società. Nè i fulmine di Giove, nè la spada di Marte, nè i baci di Venere hanno fatto tante vittime quanto Bacco coi calici spumante.*"—Bizzozero.

Alcohol, the moment it enters the cells and nervous filaments, revives their functions and excites and stimulates their action; this state of exaltation passed, however, if more alcohol is imbibed by the cells and nerves a period of exhaustion supervenes. The presence of this foreign body in the organism, tainting the blood and diffusing its vapors through the substance of the brain, interferes with the chemical processes of the body, augments the resistance to the nervous movements, and engenders that particular kind of poisoning known under the name of intoxication.

It was owing to wine that Ham was cursed and became the servant of his brothers' servants. It was owing to wine that the ancient Persians, Lacedemonians, Romans—active, vigorous, and glorious by a thousand splendid victories, as long as they possessed the virtue of sobriety—declined and fell when—

*Della stoica incude*
*Spessa nel vin tempravasi*
*La rigida virtude.*

But that was the abuse not the use of wine.

Every one should know that wine, drunk in moderation or with *temperance*, favors and augments the secretion of the gastric juices and so aids digestion; it excites the imagination, awakens the memory, dispels care, restores the physical force, and renders the movements of the body active and vigorous.

A proof of this, if one is needed, is furnished by the fact cited by all writers on hygiene, that if in the war of 1870–71 the German army was able to sustain the fatigues of the campaign and sieges, always remaining in good health, it was because they were invading and conquering a wine-producing country.

Bacchus is the "Dio salvatore." Plutarch, in the life of Cæsar, mentions that the whole army of the General was once afflicted with a disease which Cæsar cured by allowing all the soldiers to get solemnly drunk. From that day they all commenced to recover.

Certainly among the curative resources at the disposal of hygiene and medicine there is none more frequently used than wine. We always, as it were by instinct, say to a convalescent: "You should drink wine."

Hippocrates says: "Wine is a liquid marvelously adapted to man, well or ill, providing he take it at the proper time and in quantities suitable to his constitution."

Liebig, too, is of the same opinion, for he writes: "Wine is unsurpassed by any product, natural or artificial, as a restorer of the vital forces when they are exhausted; it animates and revives the saddened spirits, it serves as a corrective and antidote in all irregularities of the animal economy, which it preserves from the passing ills to which inorganic nature subjects it."

Wine considered from an alimentary point of view has its chief importance in the union of alcohol with an acid liquid; the acid moderates the too energetic action of the alcohol, especially its action on the nervous system.

The tannin and coloring matter, when present in due proportion, exercise a very favorable influence on the stomach by animating the energies of the digestive functions.

The aroma, the bouquet, the "sève" of a wine are also useful, as many facts tend to prove, among others, the fact that well-flavored substances in general have a favorable influence on nutrition.

Wine has a density nearly equal to that of water, and is absorbed into our system with much less rapidity than spirits; this fact is of great importance to the animal economy, because the effects of wine are thus felt for a longer time and without the danger accompanying the rapid effects of brandy.

Wine is absorbed by our digestive organs without any change but that of being mixed with the gastric juice. There is no need of the intervention of the digestive ferments to facilitate the absorption of the wine in its last office of nutrition. This explains its utility in certain diseases.

The complexity of the organic matters that enter into the composition of wine, which up to a certain point resembles that of the human body, explains its restorative action in the case of individuals weakened by anæmia or insufficient nourishment, etc.

Wine, then, is produced and drunk, and of all fermented beverages it is the most healthful, and the one that most harmonizes with our organism. If nature had gifted man, as it has all other animals, with a surer instinct in the choice of the food that was best suited to his constitution, certainly without any hesitation among the first substances he would have selected wine; however, having a less reliable instinct than he might have, he has allowed himself to be greatly influenced by tradition and imitation in the choice of his beverages.

# I.

## CLASSIFICATION.

Of the numerous classifications that have been made, and that might be made, of the various and diverse wines produced in the different wine-growing regions, that is to be preferred which, up to a certain point, can be considered as the most natural, by giving an immediate idea of the principal characters presented by a certain wine or category of wines.

Carpené very justly considers the classing of wines according to different dishes or repasts as misleading and hurtful to the trade; for, as he well remarks in one of his articles, if this classification should be carried out we should have tripe wine, cheese wine, macaroni wine, etc. As every one knows, the order of wines and dishes through the repast is influenced by fashion and caprice. To-morrow, perhaps fashion will oblige us to imitate northern nations and Americans in our "cuisine," and then we will be obliged to drink champagne through the whole dinner; thus champagne must be successively known as an oyster wine, a soup wine, a roast wine, and heaven knows what else.

Not long since I was at a banquet, and by chance was placed next to a certain high functionary who was to commence the series of toasts. On the appearance of the roast our high functionary prepared himself. "But how is this," he exclaimed to a neighbor, "do they not give us champagne now?" "They serve the 'roast wine' now," replied the other. "Roast wine," cried the surprised high functionary, "but at court they serve champagne with the roast." Champagne was afterwards brought, and then the eminent personage was able to get up and make his toast, a very appropriate and happy one. I cannot say what influence the "roast wine" may have had on it.

This classifying by dishes is certainly all wrong, but if we should ask ourselves the question, as an amateur does in the wine taster's *vade mecum*, "*La vite ed il vino*," "When should one drink wine?" the answer most certainly would be, "Whilst eating." Without a good selection of wines the most perfect bill of fare loses all its value.

High-class red wines should not be drunk before they have been eight or ten years in bottle. Before that they may be rough, and not particularly pleasant to the taste. Very fine white wines, too, should be well aged, otherwise the sugar, of which they contain a certain amount, will not have been all transformed into alcohol, and lessens their strength and bouquet.

A natural, primary, and main division of the various wines may be made with reference to their color, viz.:

### WHITE AND RED WINES.

It should be stated here that this general division rests not only on the color that the wine may have, or on the presence or absence of œnocyanin in its composition, but on other characteristics in which a white wine differs greatly from a red.

This division is of no little hygienic importance, wines of different color having as distinct effects on our constitution as wines of different age, alcoholicity, or acidity.

White wines, as is well known, are obtained from white grapes, or from red grapes which, instead of being crushed and fermented in a mass, are pressed, and the must fermented separately; that is, not in contact with the pomace or solid parts of the grapes.

I call attention to the fact that white wine can be made from red grapes, because wines so made have exactly the same action on our system as have white wines made from white grapes.

Certainly the following from Guyot is very true:

Wine which has been fermented in contact with the stems, skins, and seeds of the grapes is very different from that which has been fermented separately. The latter wine is white, the other red, and the antithesis, though expressed here simply by the opposition of color, does not consist in the least in this difference of color, which is only an accident. The real difference consists in the special and often opposite hygienic qualities of these two kinds of wine. Nowadays they make red wines which have all the hygienic properties of white wines, and it is possible to produce white wines which would have all the hygienic properties of red. All that is necessary to obtain this last result is to ferment the must of white grapes with the skins, seeds, and stems, in the same way as red wine is treated; in this way all the effects are obtained of a rapid decomposition and solution by maceration of the principles and products which are not found in the juice of the grape. * * *

I insist on the true distinction of wines obtained by the fermentation of the juice of the grape completely isolated from its accessories, and those made by fermentation of the juice, together with all, or at least part of the rest of the grape—a distinction quite independent of the color. Nothing is more alien or of less importance to the quality of a wine than its color. It may be a sign—an indication—but it is never a quality of itself. By the majority of consumers color is looked upon as a guarantee of the purity, quality, and strength of the wine. It is on account of this considering color as a sign of quality that unscrupulous dealers make use of it to commit innumerable frauds.

White wines are in general diffusible stimulants of the nervous system; if they are light they act rapidly on the physical organization, of which they intensify all the functions. It seems that they escape just as quickly through the skin and mucous membranes, and, above all, with the urine; their action, then, is of short duration.

Unlike white wines, red wines are tonic and persistent stimulants of the nerves, the muscles, and the digestive organs. Their organic action being slower is more prolonged; they do not unduly excite the perspiration nor the excretions, and their general action is astringent, persistent, and concentrated.

Moreover, the common opinion, founded on daily experience, leaves no doubt of the real difference, in their sensual and organic effects, between white wines and red.

## Of equal importance are the following words of Dr. Gauber:

If one should divide the grapes gathered from a vineyard of the "Graves" of the Gironde into two parts, and of one make white wine and of the other red, and then, at the end of four years, make a careful tasting of these two wines which have been carefully treated during these four years, what will be the result? Made from a raw material apparently identical, will they be equally developed and equally mature? The white wine will have aged the most.

Will they produce the same effect, the same degree of stimulation, on our organs? Let us collect the sensations produced by one and the other in the order in which they are produced.

1. A glass of white wine, well made and dry, the moment it enters the mouth develops a bright and penetrating aroma, and leaves, in passing, an impression, agreeable it is true, but fugitive and almost hot. Hardly has it reached the surface of the stomach when it causes a feeling of warmth which, in less than ten minutes in the case of certain healthy but impressionable constitutions, becomes very intense. Sometimes the action, by sympathetic radiation, is reflected from the stomach to the head with the promptitude of the electric fluid. Generally, after an hour or less, a sensation is felt as of a pressure either on the two temples or around the whole head; the hand is instinctively passed over the forehead as though to free it from some load. Sometimes a feeling of painful fullness of the brain accompanies these effects. The irritation is communicated from the gastric and nervous centers to the whole body. It shows itself by increased warmth, often irregularly distributed, of the body (with irritable people the palm of the hand often becomes unpleasantly hot and dry); by a need of movement, of displacement rather than of exercise (with people of the disposition mentioned above this need is shown by an internal agitation, by slight muscular tremblings accompanied by shooting pains that strike, with the rapidity of lightning, different parts of the body).

At the end of two or three hours, more or less, according to the temperament and susceptibility of the individual, the irritation passes away and the taster finds himself in the same condition as before, with or without a certain feeling of lassitude or sadness.

2. If the white wine is replaced by a red wine of the same vintage, and taken at a proper temperature, it will leave in passing a distinct impression on the two senses of smell and taste of a soft aroma; its fluidity in the mouth is less, and though it feels more material, so to speak, it leaves a less intense feeling of dry heat. Its contact with the stomach produces a softer and more gradual impression.

The organ is still warmed, but in a more vital manner, as it were. As to the sympathetic propagation of the stimulating action towards the head, it still takes place, but without the nervous phenomena of pressure and pain; the brain is gently excited. Its extension to the organs of the senses, if it takes place, is no longer betrayed by the need of displacement and agitation, but by a strengthened desire for exercise, which is very different. The duration of the stimulation is more prolonged and ceases insensibly, so that the most attentive observation cannot detect the exact time at which it ends.

Here is, we believe, the sufficient explanation of the difference of effect observed between white wine and red wine—the first (white wines of Graves), produced by fermentation of the must separate from the pomace, contains about 4 to 6 per cent of extractive matter and tannin; the second, 8 to 11 and 12 per cent of the same matters.

It is to this difference in the proportions of the rough and astringent matters of the wines that we attribute their different effects.

In the red wines the pressure of the alcohol on the nervous system of the stomach is softened by the interposition of more abundant tonic and extractive matters; the effect is thus slow and successive. In white wines it is almost immediate, and therefore stronger and less lasting.

Each of these large groups into which the various wines may be divided is susceptible of three subdivisions, which are sufficiently natural, as they give immediately some idea of the quality of a wine which enters into any one of them.

These three subdivisions are the following:

1. Table wines.
2. Dessert or alcoholic wines.
3. Blending or cutting wines.

### 1. *Table Wines.*

These wines may be of higher or lower quality, according to the locality in which they are produced, and to the care that is taken in their making and after-treatment; they must not be sweet nor too alcoholic; not aromatic nor possessed of too pronounced a bouquet, though those of higher quality may be slightly aromatic; they must not be too rich in color, too astringent, nor too acid; they ought not to be harsh nor of too heavy body, that is, too rich in extractive matter.* A wine of this group should be clean tasting, and should form an harmonious whole, agreeable to the palate and stomach, so that it can be drunk with pleasure. These wines are healthful, because they favor digestion, and a certain quantity of them can be taken without producing intoxication or other physical disturbance.

Concisely the characters of a typical table wine may be described as follows:

Light but not poor in alcohol; not the slightest tendency to sweetness; pleasing but light and delicate aroma and flavor; nothing excessive, but complete harmony of all parts. A full and generous homogeneity; limpidity; constancy of type. Though in the matter of dishes variety

---

* "In the middle of the seventeenth century England consumed the light wines of France, and, as Gladstone says, they laughed and sang in those days in the British realm. The wars between France and Great Britain breaking out, the French wines were prohibited and in their stead the heavy wines of Spain and Portugal were imported; they still drank as much, continues Gladstone, but they sang no longer; to laughter succeeded quarrels and base deeds."—R. Dejermon.

is both useful and pleasing, it is different with wine where constant
uniformity of type is necessary.

As in this class of wines are comprehended all qualities from the
finest to the most ordinary, it is easily seen that other distinctions can
and must be made, in order that the wines, for example, of Barolo or
Chianti, shall be distinguished from wines produced in some less favorable locality.

The various wines that enter into the category under discussion can
be naturally and conveniently classified as follows:

A.—Superfine, or high-class wines; the "Grands Vins" of the French.

B.—Fine wines.

C.—Fine common wines.

D.—Common wines.

E.—Low-grade wines.

This classification, as Polacci would say, has nothing imaginative
or strained about it, as it simply represents the wines that we really
have and of which we make use in commerce.

I will now try to give, not a definition, because the name of each class
is of itself a definition, and should give a fair conception of the distinction to be made between the several classes; but an idea regarding
the characteristics which have served in grading the wines which we
actually produce in Italy.

A. *High-class Wines.*—These are wines which are produced in certain
spots, or rather which are obtained from certain varieties of grapes, grown
in especially favorable conditions of climate, and more particularly of
soil, compared with those of the circumjacent vineyards; wines which
also, it may be said, are the product of an almost infinite series of careful treatments, beginning in the vineyard and continued through the
vintage and during the whole time, which is certainly not brief, of their
conservation; wines, in short, which unite in themselves all the characteristics and qualities which should be found in a fine wine, united
with the greatest delicacy and fragrance of aroma and freshness on the
palate. An Italian wine which belongs to this class is the Chianti di
Brolio. Of the French wines of Bordeaux, or more precisely of the
Médoc, there are Chateau-Lafite and Chateau-la-Tour, the latter of
which is distinguished from the former by a slightly heavier body and
a more pronounced flavor and aroma.

B. *Fine Wines.*—These are wines which approach very nearly to the
preceding class, but are, nevertheless, somewhat inferior to them, either
in delicacy of aroma or in some other quality; very often they lack or
are deficient in the freshness which distinguishes the first class. These
wines are very often the product of grapes grown in the neighborhood
of the vineyards producing the first-class wines which have given
renown to the locality, but they may be made from grapes grown in
other localities. To this second class belong, for example, those wines
of Chianti which resemble greatly in character the Chianti di Brolio,
but do not equal it. In the same way among the French wines of the
Médoc, Saint-Julien and Saint-Estephe approach but are not equal to
Chateau-Lafite.

It may very possibly be that some of the wines of Chianti exhibit
qualities which place them, so to speak, in rank with the Chianti di
Brolio; then from the second they must be promoted to the first class,
as is the case with Chateau-la-Tour, which, though somewhat different, is

deemed worthy to stand in rank with Chateau-Lafite and the other two, Chateaux-Margaux and Chateau-Haut-Brion, which together form the four "grands vins," high-class wines of the Gironde.

'C. *Fine Common Wines.*—In this third category are placed those wines which are intermediate between the fine wines and the common wines. This class of wines can be produced in large quantities in Italy, as there are numerous regions both in the hills and plains which present the requisite favorable conditions.

The wines in question generally lack or are deficient in delicacy; with time, and sometimes, too, with a little artificial aid, they acquire some aroma which is not, however, always very delicate. These wines form, or ought to form, the bulk of our export trade; but if we wish to do a steady trade we must set ourselves diligently to make and properly handle these wines. To do this the producers must abandon the idea of making high-class wines, and confine themselves to wines of this kind.

The wines of this class produced in Italy, especially by those who have recourse to artificial additions, or who do not well understand the processes of wine making, present a certain dryness to the taste which is not exactly pleasing.

The taster will pronounce them sound wines without any particular defect, but he is not quite satisfied. This may be owing to an artificial aroma, or to the addition of alcohol; it may be caused by heating, or by a too violent fermentation, to the grapes having been picked at the wrong time, or to an injudicious correction of the must, or—but as this is not the place to try to account for it it will suffice to state the fact.

Such artificial aids, then, as the addition of drugs, the drying of the grapes, heating, etc., should be abandoned, and instead a judicious choice of vines, or a blending of grapes or wines substituted; in this way it will be possible to deliver to the trade wines which have a sufficient freshness of taste and frankness of flavor; they will be to a certain extent smooth and delicate, and will possess more or less of that fruity taste so much liked by consumers.

D. *Common Wines, or Wines of the Plains.*—This is a class of wines of which it is not very easy to give a definition or to point out its exact limits in order that it may not be confused with the preceding or comprehended in the following class. To prove that this is a real difficulty it will suffice to quote the eminent agriculturist, F. Re: "I have sometimes drunk wines made from grapes grown in a naturally clayey soil, subjected to irrigation, which were very good, and some even which seemed to be of superior excellence."

I should therefore state that all wines grown on level ground cannot be classed as common wines; even on the plains, when the climate and especially when the soil and the variety of grapes are particularly favorable, choice wines may be produced which are worthy to figure in the preceding class.

The division or class of common wines comprises all those wines consumed in the largest quantities, and which, because of the ease and economy with which they are produced, can be sold at a low price, so that they find steady consumers among the working classes, who consume, after all, the greater part of the product of the vineyards.

These wines are most commonly the product of grapes grown on the plain, either in vineyards or associated with other crops; this does not exclude the possibility of producing such wines from grapes grown on hills, and especially when the exposure is unfavorable, or when the nature of the soil is unsuitable, or when, on account of the ignorance of the grape grower, who prefers quantity to quality, he plants by preference those varieties which give an abundant crop of very inferior grapes. Wines of this class have very poor keeping qualities, lasting two years at the most, and in general in aging, with the exception of those which are very rough and astringent, deteriorate instead of improving.

. These wines are sufficiently alcoholic, but owe their conservation less to their alcohol than to their acids, among which, with many of them, must be included carbonic acid. To their acids, also, they owe most of their hygienic value, which is to aid in the digestion of the food consumed by the laborers who drink them—food which is naturally difficult of digestion, and rendered more so by its ill preparation.

These wines are more nutritious than are those of the preceding class, containing, as they do, larger quantities of albuminoids, in which grapes from the plains usually abound. The reason of the greater abundance of nitrogenous matters in inferior grapes is the natural fertility of the soil on which they have been grown, or the fact that this ground has been manured with nitrogenous fertilizers, with the idea of increasing the bearing of grapes or the production of wood and foliage.

These wines are naturally very variable, differing greatly according to the conditions of soil, climate, and aspect under which they have been produced. To further increase this variability man does his best, seeming to take a delight in practicing methods of wine making that are apparently ingeniously calculated to spoil the wine.

A wine of this class should be of easy digestion, and easily consumed in moderate quantities, without affecting the head or the stomach. It should be smooth, clean tasting, well fermented, with a certain amount of flavor and acid, and should show none of the effects of secondary fermentations to which these wines are so subject; finally, it should possess a good, bright, but not deep, color.

I have said a wine of this class "should be" all this, because only too often, on account of careless making or improper handling, they are anything but healthful; they are, on the contrary, heavy and indigestible, causing, even when used sparingly, disturbances of the head and stomach; they are heavy-bodied wines, and so thick as to be appropriately called by some people, "vini carnosi;" their defects are usually due to the vessels in which they have been made and kept, to bad fermentation, or to the addition of substances which have been put in with the intention of preserving the wine, or of masking its defects. They are often costive and overcharged with tannin and coloring matter, recalling, the moment they touch the palate, the flavor of ink. Their color is generally unstable and dull.

E.  *Low-grade Wines.*—These wines occupy the lowest grade on the œnological scale, that is to say, among natural wines. In drinking one of these wines one asks himself if it is really a wine or not rather a piquette or mixture of water and wine, with superabundance of the former. Except color, these wines are deficient in all the elements proper to wine. They must be consumed promptly during the winter, or they cease to be wine. Generally, to render them drinkable at all, they must be left for some time on their pomace, or on that of better wines; or else they can be cut with other wines, or be given the treatment usual in Tuscany, known as the "governo."

When these wines are sound they do very well for cutting with other wines, thus making a blend which can be classed with the common wines, or even sometimes with the third class, or fine common wines.

## 2. *Dessert or Alcoholic Wines.*

This class includes all those wines which the French call "vins de luxe," and therefore champagnes and other sparkling wines, which, however, are, unlike most of this class, of relatively low alcoholic strength.

Sparkling wines are placed here because, as a rule, they are of high cost, and therefore "vins de luxe." However, we are now producing natural wines which arè afterwards artificially made sparkling, at a much less cost; and this industry is assuming such proportions that it cannot well be overlooked.

Apparatus of different kinds for the production of sparkling wines have been known and used for a long time in France, Germany, and Austria.

Latterly the practice of artificially making champagne from natural dry wines has been extensively followed in Italy; this is due to the invention of the apparatus of Carpené, which possesses above all previous systems the advantages of simplicity and cheapness. This system has rendered possible the production of good sparkling wines at a moderate cost.

With this explanation regarding champagne, and the reason for placing it in this class, I pass to those wines more properly belonging to it, and here give Polacci's definition of "vini di lusso."

These wines are nearly always alcoholic, more or less aromatic, and are drunk, as a rule, after dinner, on which account they are called by foreigners dessert or after-dinner wines. They are, so to speak, concentrated, and are sipped from small glasses like cordials, for which reason the French know them as "vins de liqueurs." We know them as "vini di lusso," because they are certainly not necessary beverages, and from their high cost are usually reserved for the tables of the rich.

The many and diverse wines of this class can be divided, or rather united, under the following heads: Sweet Wines; Alcoholic Wines; Sparkling Wines.

In this class are wines so well known, and of such special character, that it is difficult to class them together, and each is usually spoken of by itself as almost forming a class apart; as with the wines in the first class, the "grand vins," their qualities and peculiarities are so well known that their names alone is a sufficient description; such wines are Marsala, Lacrima Christi, Vernaccia di Sardegna, Malvasia di Lipari, etc.

## 3. *Cutting Wines.*

These wines are rich in alcohol, coloring matter, and body, but often deficient in acid; they cannot be drunk alone, and the only reason for producing them is that there are localities which produce wines which are thin, poor in color, weak in alcohol, and generally lacking in those qualities which wines of this class have in excess. A mixture of these two kinds of wine, each of which alone is of little value, produce a wine which is sustaining and nutritious, and especially suited to the needs and means of the laboring classes. The better kinds of these wines may even be blended to form a wine which might be placed among the fine common wines, or third class, and not unworthy of the honor of bottling.

At the present day the French wine merchants use large quantities of cutting wines imported from Italy, Spain, and Dalmatia. Before the invasion of the phylloxera, their blends were made with the wines of Roussillon, Languedoc, Pyrenées-orientales, Aude, Gard, Tarn, etc., all wines rich in coloring matter and alcohol, solid and heavy-bodied, and at the same time smooth, delicate, and with a characteristic and persistent aroma which is very pleasing.

Here is, for example, a blend or mixture of different wines formerly much in vogue in France:

| | |
|---|---|
| Wine of Roussillon | 30 litres. |
| Wine of Narbonne | 60 litres. |
| Wine of Cher | 30 litres. |
| Wine of Poitou, blanc | 60 litres. |
| Wine of Bourgogne | 30 litres. |
| Wine of Pique-poule, at 15 per cent | 15 litres. |
| Total | 225 litres. |

A French writer thus justly expresses himself: "After the invasion of France by the phylloxera, commerce drew contributions from all wine-producing regions; science was also brought to its aid; an immense productive movement commenced, not only in France, but in foreign countries, and now wines flow in from all parts, from Spain, Italy, Austria, Greece, the Crimea, and even from Australia; wines of all kinds, which, passing through the skillful hands of our merchants, there receive the official seal, the inimitable touch, which serves them for passport to the wine connoisseurs of the entire world." Further on we read: "In this combination each region plays its role, and helps towards the final result that we desire to obtain; from Italy the blend obtains strength, extract, body; Spain supplies softness and fruitiness; our own wines add piquancy, and economize on the price of production."

In whatever way the cutting is done, and whatever the combination adopted, the following may be taken in general as the composition of most blends:

One third wine of Italy;
One third wine of Spain;
One third "petits vins" of France, or wine made from dried grapes.

Cutting wines are then of no little importance to wine growing in France, or rather to the French wine trade; why then, should they not be as important to ours, especially now that the two are in competition?

Let us then produce cutting wines, but let them be well made and sound. By such wines the Italian wine trade will be benefited as much as is the French now.

## II.

### TASTING.

The word "tasting" is not used with its ordinary signification when referring to wine, but means, in that case, not only the testing of its flavor by means of the gustatory organs, but also a careful examination of the wine in other ways; of its appearance, of its bouquet, as well as of its effect upon the palate; all of which is necessary before a final judgment can be passed on its character, its qualities, and its defects.

Wine tasting is a somewhat difficult art, which cannot be acquired without long practice, and then only by one who possesses a clear eye and very delicate organs of taste and smell. When the last two organs have the requisite sensibility, practice alone is necessary to give them the skill needed in tasting a wine.

It is by frequent tasting, by making comparisons, by the examination of good types, that this delicacy and sensibility of the palate is developed which enables it to detect and appreciate the faintest aroma, flavor, or bouquet, as well as the slightest defect.

Practically the tasting of a wine is, up to a certain point, of more importance than its chemical analysis. Analysis shows us the principal components of the wine and the proportions in which they are combined; tasting tells us whether these components are in proper proportions to form an harmonious whole, or are, some of them, in excessive or deficient amounts; whether the wine has " sève," bouquet, aroma; whether it is mature or not; whether it should be racked or bottled; what its defects are, its keeping qualities, etc.

Any one can say whether a wine pleases him or not, but only the experienced taster can pronounce with any degree of certainty on the real properties and character of a wine. A good wine may be pleasing to-day and not so to-morrow, on account of slight exterior influences which are dangerous to its stability but may be only transitory in their effects, and the wine may recover and be as good as ever.

In order to make useful deductions it is of the highest importance, in fact absolutely necessary, to be able to appreciate and reflect on the sensations experienced in the tasting. It is not every one who can appreciate the true import of what they perceive, but only those who have trained themselves by long practice.

The experienced taster, when called upon to give his opinion, looks at and attentively examines the wine. He then agitates it by shaking the glass, and, when necessary, places his hand round the glass in such a way as to warm the wine, thus favoring the volatilization of those matters which affect the olfactory organs; he then tastes it.

Sometimes the simple agitation of the wine by twirling the glass is not sufficient, especially when the sparkling and bouquet are to be particularly noticed. In this case the wine must be more thoroughly shaken, which is done by placing the palm of the left hand over the mouth of the glass, and then striking the bottom of it forcibly against

2

the knee. This causes the wine to give off its odors, and in the case of sparkling wines its carbonic acid, more freely. The method, writes Ottavi, is not very polished or elegant, but accomplishes the purpose very well.

As can be easily seen the wine taster should preserve his senses, that is, those of smell and taste, with their utmost sensibility; this is only done by avoiding excesses of all kinds, for these in course of time are bound to diminish that sensibility, or to destroy it completely. Thus he must abstain from all highly alcoholic beverages, from strongly salted or flavored dishes, from tobacco in any form, and in general from everything that acts too energetically on the organs of the above-mentioned senses.

Physical indisposition, more especially affections of the nasal organs, the mouth, or throat, diminish or destroy all sensibility of the senses of taste and smell.

"Wine should not be tasted fasting, or it will taste weak and insipid; nor after drinking wine; nor with a full stomach. Moreover, the taster should not have eaten anything sour, salt, or bitter, nor anything which might change his taste; but he should have eaten a little, but not yet have digested it."—Carlo Stefano.

The taster should not attempt to give his opinion of more than a certain number of wines at a time, as after having tasted a certain number the senses become temporarily much impaired and incapable of nice discrimination; nor should he judge of a wine after an abundant repast, as the various flavors of highly seasoned or sweetened foods have a great influence on the palate, and prevent it from judging a wine critically.

It is a well-known fact that after eating sweet fruit a wine seems to be rougher and harsher than it really is, whilst cheese, nuts, artichokes, etc., make it appear smoother and more delicate.

With piquant cheese, like Parmigiano and Roquefort more especially, which Grimod de la Reynière has called "the tippler's biscuit," all wines seem good, or at least much better than they really are. It is also true that strong and badly tasting wines when drunk undiluted destroy the sensibility of the palate; people habituated to these wines end by being unable to find any taste in the fine wines of delicate flavor which are the delight of the connoisseur.

Tasters who are accustomed only to high-class wines, when they taste ordinary or low-class wines are apt to underrate them, if they do not reject them as altogether valueless, though they may be sound and clean tasting.

On the other hand, tasters accustomed to ordinary wines almost always deem the prices paid for high-class wines excessive.

This suggests the importance of habit as a factor in the modifications which the taste may undergo. It may easily happen that the prolonged use of a substance may render the sense of taste obtuse, and that the tongue may become "saturated," as Brillat-Savarin says in one of his happy aphorisms. Thus, when the palate has become habituated to a taste, that which at first was intolerable becomes often pleasing and even necessary. Generally, however, habit educates the sense of taste and renders it acute.

Sometimes a taster is called upon to give an opinion as to the character, the good or bad qualities of a wine of a certain locality or of some particular producer or vineyard; in this case, even though he may be well acquainted with the kind of wine, to be able to give his decision with more confidence, he will carefully provide himself with a wine of the same type as that which he is called upon to judge; he can thus receive material aid by making a comparison.

Naturally, a taster who is used to the wines of a certain locality or country will be more easily able to detect the slightest differences between the wines of that locality, especially those differences in fine wines which distinguish wines produced by different vineyards even in the same locality, and when planted with the same varieties of grapes.

A taster should be very cautious in giving an opinion of a young wine, or of one whose origin is unknown, and of pronouncing on its intrinsic worth; the youth of the wine will often mask defects, which, later, become apparent.

When it is found necessary to taste several wines in succession, it is a good practice to eat a little dry bread between each wine, or to rinse out the mouth with a little fresh water, to neutralize the palate, so to speak.

It is always good to rinse out the mouth with fresh water before commencing to taste.

Before commencing the tasting, or rather the final tasting—that on which is based the concluding judgment—the wines should be sorted; for example, if the wines are of the same kind, but of different ages, it is best to begin by tasting the weakest, thinnest, or greenest wines, reserving the maturer wines and those which are more aromatic, smooth, or alcoholic for the last.

The same is true when there are many and diverse wines, as at an exposition. In this case the tasting proper should be preceded by an arranging of the various wines, a thing which is not done at all, or badly done as a rule, much to the detriment of the exhibitors. This selection should be based not on the labels on the bottles, or on the statements of the exhibitors, but on a preliminary tasting; in this way those who are to judge the wines will not be presented successively with different types of wine, with wines of different qualities and ages together, and, as is unfortunately the case, sometimes with defective or bad ones.

There are tasters who are ready at any time to pass judgment on a wine; they will even taste directly after smoking. Their opinion, to say the least, is of little value.

A good taster is not always in condition to exercise his art, and for that reason must sometimes refuse to make a tasting when he does not feel in a state to judge critically.

Here I may appropriately remark that the wine dealer often relies too much on the lack of delicacy of taste on the part of the consumer. He should remember that among his customers there is occasionally a connoisseur, or at least a fairly good taster, who can appreciate the wine at its true worth, and whose opinion is followed by the majority of his other customers.

A little advice is needed also by those who are called upon to judge competing wines at exhibitions or elsewhere.

Without exaggeration, I may say that there is scarcely a person in

Italy, connected in any way with wine, who has not been called upon to act as judge in competitions of this kind. I need not say how much harm this has done our national wine industry; I will simply, with Polacci, express the desire that we might see some day in Italy "una vera magistratura enologica," a body of competent men to look after these affairs.

We will now return to our tasting. The forenoon is the time best adapted for wine tasting; the wines are of the proper temperature, a temperature which varies for red wines between 54° and 60° F., and for white wines between 50° and 54° F.; the taster is in good condition, and consequently the tasting may begin.

There should be no bad odors present, and the place in which the tasting* takes place should be well lighted with diffused light, not obscurely through a small and narrow window, nor too brightly by the direct rays of the sun; it should be remote from all noise, where the taster can remain quite undisturbed.

It is a fact admitted by physiologists that the senses exercise a mutual influence on one another, so that anything that excites one sense has the effect of increasing the acuteness of the other.

This reciprocal influence seems to be confirmed by the recent researches of Dr. Albertini, who says that the defect of color-blindness is accompanied by a corresponding deafness for certain sounds. Thus, those who cannot perceive red cannot distinguish *sol*, while those who are color-blind for green are unable to recognize *re;* to this lack of oral perception is joined the inability to reproduce these notes with the vocal organs.

"The taster," writes Franck, "should be deaf and dumb; deaf, in order that his judgment of the various qualities and defects revealed to him by his senses may be undisturbed; dumb, in order to prevent the expression of a hasty or insufficiently considered opinion."

Every one has noticed how a gourmand will close his eyes in order better to appreciate the delicate flavors of a substance, thus bringing his mind to a proper state of attention by the absence of all other excitement. This will explain the exclamation of the court parasite, who, disgusted with his too turbulent table companions, cried: "Hush! You do not understand what you are eating."

---

* Here the question asked in "Conseils d'un amateur:" How should wine be drunk? might appropriately be answered. In our opinion, in order that the benefits of drinking it may be enjoyed in their fullness, the first thing necessary is that the wine shall be presented in the manner most pleasing to the eye and to the palate, for this impression on the senses has a most important influence on the rest of our body. With this end in view we should be scrupulously careful to have the wine at the exact degree of temperature that the nature and quality of the wine demand for the proper development of its flavor and bouquet, and then to make a judicious choice of the kind of glasses in which it is to be served. For Bordeaux, Burgundy, Chianti, Barolo, etc., the proper temperature is that of the dining-room, where they should be placed for some hours before they are to be consumed. White wines, sweet wines, etc., must be of the temperature of the cellar, that is, supposing the cellar is very cool, otherwise it is necessary to cool the wine, either by placing the bottles *on* ice, or by placing them in water containing a few lumps of ice, but never *in* the ice, for that completely destroys the character of the wine. Champagne is the only wine that may be put in ice, but even in this case discretion should be used, and if the wine is put in ice for three or four hours before being used it will be found sufficient, and the wine should then be served directly from the bottle. It is then a great mistake to place wine in ice or in freezing mixtures, for a wine so treated destroys the appetite and is injurious to the health.

The practice of pouring champagne into decanters containing ice cannot be too strongly deprecated. In the first place, it is not wine you drink, but a mixture of champagne and water; and secondly, the temperature is never right, as it cannot be regulated.

Let us add that ice should never be put *into* wine, for it destroys the bouquet and flavor of the wine, and if it gives a momentary pleasure to the palate by a sense of coolness, it also renders the digestion slow and laborious.

The taster should be provided with a porcelain cup, or with the Bordelais silver cup, which, however, may be made smooth, and if so, the bottom should be a little raised; this cup is especially applicable to young or blending wines, as it is the best for observing the tint and intensity of color and the degree of limpidity.

There are two kinds of Bordelais cups; one preferred by the sellers, and the other by the buyers.

Naturally the seller tries to show off his wine to the best advantage; for this purpose he prefers a cup with a raised bottom, bright, shining hollows in the sides, and a large rim, on which the rays of light have a pleasing effect.

The high rim and the yellowish tint that the maker gives to the silver of the cup concur to improve the appearance of the wine. The buyer's cup, on the contrary, is of silver of its natural color, and without the exaggerated rim, and without anything that might modify the appearance of the product to be examined.

In Bordeaux they prefer a cup almost without border, a kind of plain saucer, having in the center a slight convexity. In this cup the wine appears exactly as it is, without the slightest artificial alteration.

Lately the buyers of the Gironde have begun to use the twin cup—that is, two cups joined together with a hinge—by means of which it is possible to have two wines, which it is desired to compare, in almost the same conditions with regard to light.

Besides the Bordelais cup he should have at his disposal glasses of various forms, but all thin and homogeneous. Some should be chalice-shaped, but not too long; some of the shape known as "Bordelais;" some cognac glasses, narrow at the mouth and widening below, that is, truncate egg-shaped. By means of the latter, the bouquet, fragrance, and odors generally can be best perceived, especially when their disengagement is aided by shaking.

Conical glasses, on account of their form, serve very well to judge of the color of a wine, as according to the height in the glass where the wine is examined, there will be a greater or less thickness for the rays of light to traverse. Between the two extremes the differences of tint (the gamut of color going from rose to red in the case of red wines, and from white to golden in the case of white wines) is very interesting, and may sometimes give very useful hints.

The different aspects under which a wine can be considered are so numerous, there is such an almost infinite number of possible differences in the various qualities and defects that have to be considered, that even the most expert taster would find himself in great perplexity without a proper and systematic arrangement of his sensations. To avoid this perplexity he proceeds as follows:

He takes a glass containing a small quantity of the wine; raises it to a level with his eyes, examining it carefully first at arm's length, and afterwards more closely; raises and lowers the glass in order to view the wine from above and from below. By inclining the glass and viewing it in different positions, by giving the wine a rotary motion, making it rise up the sides of the glass, he is assisted in his observations. In this way the taster learns all that can be discerned by the organ of sight, namely: the color or colors, the degree of limpidity, the disengagement of bubbles of gas, and the degree of persistence with which they cling to the sides of the glass.

Its appearance is, to a certain point, a sign of the condition of the wine; from it the taster receives his first impressions and begins to form his opinion; this opinion is as yet, however, very relative, and rests only on probabilities, as a good wine may possibly wear the aspect of a bad one.

"Limpidity and vivid color are favorable signs," writes Guyot, "but they do not constitute high quality, though the contrary appearances are real defects."

Thus, though the eye may be pleased, the nose and palate may not be.

The experienced taster will be able to tell, to a certain extent, whether the color is natural and homogeneous, and so to a certain extent whether it is artificial; in this latter case he will be able to make a probable guess at the nature, vegetable or mineral, of the substances used to give color to, or to enhance the color of, the wine.

The estimation of the color of wine is very important, especially with cutting wines which are to be mixed with others to obtain the type demanded by customers.

The eye having fulfilled its office, it is the turn of the olfactory organs.

The sense of smell resides in the ample nasal cavities, and more especially in the pituitary, the mucous membrane which lines them. Odors, or better, infinitesimal particles of substance, reach this membrane by means of the external organs of the olfactory apparatus, that is, by the nostrils; they may also enter by the internal nostrils, the two openings which put the nasal cavities in communication with the larynx.

Physiologists admit that the sense of smell is not provoked only during inspiration but also during expiration, though in the latter case much more weakly. Thus, Franck tells us that it is during expiration that we analyze the perfumes of wines.

Besides the expiratory movements that we execute, sometimes quickly and intermittently, sometimes slowly, in order to place fresh portions of air in contact with the mucous membrane, the cavities formed by the folds of the mucous membrane are of great aid in the perception of odors, as the air laden with odorous particles accumulates in them, and thus prolongs the impression. The mucous membrane may be more or less sensitive according to its relative state of dryness or humidity, which, as I have shown, are much affected by colds in the head. When too dry the cellules are almost indurated, and when too moist they are separated from the air by a watery layer which prevents their regular action.

As may be supposed from the foregoing, the sense of smell will receive two impressions, or rather, will receive impressions at two different times, the first before the wine is tasted, and the second when the tongue and palate have almost finished their action; that is, when the taster commences to swallow the wine.

The sensations received the second time are various and very different from those received at first.

The first sensations are those caused by the readily volatile substances that the wine contains, and which are given off at the ordinary temperature of the wine, and without other assistance than the shaking and motion given to it by the hand of the taster.

The second series, which is perceived during or after swallowing the wine, is caused by the substances which are volatilized by the increased temperature due to the heat of the mouth and to the wine being well "subdivided" by the tongue, and finally to the action of the juices secreted by the various parts of the mouth.

The taster having thoroughly examined the appearance of the wine, lifts the glass to a convenient distance and inhales the odors which are given off, and which fill the upper part of the glass, sometimes shaking or striking the glass to aid their giving off.
A wine may give off various odors, good or bad. I will treat of both of these when I come to describe the qualities good and bad which a wine may present.

Before proceeding further with the tasting it will be interesting to repeat the observations of Guyot, and of Brillat-Savarin, the "modern epicure," regarding the colors and aromas of wines.
"The aroma, like the color," writes Guyot, "is a favorable or an unfavorable, an agreeable or a disagreeable sign; but wine is above all an alimentary beverage; it is well that sight and smell should be satisfied, but it would be puerile and ridiculous to give undue importance to the satisfaction of these two senses, and to found the pretensions of a wine to superiority exclusively on its pleasing effect on one or both of them.
"I make this remark expressly because there are many hosts who have a troublesome habit of insisting that their guests shall continually inhale the odors given off by their wine, and especially insist on their smelling their empty glasses during a great part of the dinner, at the risk of making them die of thirst.*
"The connoisseur, like the taster, knows perfectly well the importance of the color and bouquet of a wine, but he knows also that their appreciation should be immediately followed by the introduction of the liquid into the anterior portion of the mouth.
"The color and the bouquet are two introductory notes of a gastronomic theme. Alone they have but a relative value, and give but a partial impression of the whole theme."
Brillat-Savarin, who is an authority in matters of taste, writes, in his "Physiologie du Goût:"
"For my part I am not only persuaded that without the sense of smell there is no complete tasting, but I am tempted to believe that taste and smell constitute but one sense, of which the mouth is the laboratory and the nose the chimney; or to speak more literally, of which the former serves to taste the tangible parts and the latter the gaseous."
Thus, for example, when we eat a peach, the first thing that strikes us is its perfume; when we place it in the mouth we experience a sensation of coolness and acidity which invites us to continue; but it is only when the mouthful is swallowed, when it passes beneath the nasal cavities, that we perceive the perfume, and the peach completes the impression that it should produce. This will explain why the sensations which are usually accredited to the sense of taste are in reality much more complicated than is supposed, and that touch and smell contribute in great part to the complex effect. It may be said that without smell taste would be reduced to very little and its agreeable sensations much enfeebled. Taste and smell combine with and complete each other, and Thomson has very justly defined them as the instruments of a unique

---

* Here Guyot might safely add that these people who are so troublesomely importunate are generally those who have recourse to the addition of artificial aromas to their wines.

sense. It is a well-known fact that if the nose be held whilst tasting a substance we perceive the fundamental tastes, such as sweetness, bitterness, salt, and acid, but all the delicate flavors disappear completely.

We have now arrived at the sense of taste, or, as some call it, the tasting proper. The sense of taste, with its somewhat complicated apparatus, is the one which has the most important office to fulfill; by it we decide whether the wine has the freshness, solidity, strength, delicacy, etc., in short, the qualities required by the most critical taster.

The principal seat of the sense of taste is the tongue, although it seems to have been proved that both the anterior face of the membrane of the palate and the posterior part of the palate are capable of receiving gustatory impressions.

According to the illustrious Professor Lussana, the tip of the tongue is distinguished by its ability to detect the finest gradations of flavor, whilst the posterior part, on the other hand, is distinguished by the intensity of its sensations, and is therefore more impressed by repugnant flavors.

Different parts of the organs of taste receive different impressions from the same sapid substance.

The action of sapid substances in contact with the tasting apparatus is somewhat complex, and is physico-chemical rather than mechanical, as formerly supposed.

For this reason the particular gustatory sensation due to any alimentary substance is felt more keenly when the substance is kept for some time in contact with the tasting membranes, as is the case, for instance, in slow mastication.

This time, however, should not be too much prolonged in tasting wine, or it becomes impossible to distinguish between the many and diverse flavors which a wine presents.

The taster, having now critically examined the wine to the best of his ability, by means of the eye and the exterior part of the organ of smell, must pass quickly to the domain of the sense of taste.

To this end, he slightly lowers his head, carries the glass to his lips, and introduces a sip of the wine into the anterior part of his mouth, where the sense of taste receives its first impressions.

The taster retains the wine in this part of the mouth for a certain time; and in order better to perceive the various flavors that affect this part of the tasting apparatus, he divides and subdivides the wine with the tip of his tongue, or as experts express it, he "breaks up" the wine, in order to increase the surface of contact between the wine and the gums, palate, and tip of the tongue.

As soon as the taster has received a distinct impression of all the sensations caused by the wine in this part of the mouth—that is, of those due to sugar, acid, tannin, etc.—he slowly raises his head, thus allowing the wine to pass to the posterior part of the mouth, when he takes a short breath and slightly gargles; at this stage of the operation he will perceive any earthy, bitter, or mawkish taste, or any taste of wood, cork, etc., that the wine may have; here he will also remark the alcoholic strength or weakness of the wine. The wine is then, so to speak, left to itself and passes into the larynx, the œsophagus, and on into the stomach.

As the wine passes down the throat it gives off odors which, as has been mentioned, ascend to the palate and the internal nasal ducts. The effect of these odors, and therefore of the qualities and defects of the wine, is intensified if the moment the wine is swallowed the mouth is moved as though masticating something.

It has been attempted to measure the duration of certain sensations; i. e., those due to the aromas, bouquets, flavors, alcoholic strength, and the various tastes of wine.

In general these sensations are perceived in the brief space of time of 3 seconds, and their duration varies from 10 to 20 seconds. After the wine has been swallowed all the sensations disappear in about 7 or 8 seconds. In certain special cases the aromas leave a more lasting impression; bad tastes persist longer than good ones. In some wines the aroma can be perceived for 55 or 60 seconds.

The sensation due to astringency is of short duration in fine wines, and is much less intense than in the case of wines made from immature grapes, where it makes a violent impression on the lips and the sides of the mouth, which lasts sometimes for 100 to 110 seconds.

Different bad tastes have different ways of showing themselves; some are noticeable the moment the wine enters the mouth, while others are not perceived till some seconds after the wine is swallowed.

Some moldy tastes do not manifest themselves for 7 or 8 seconds after the wine has left the mouth, but persist for 100 or 140 seconds.

The "goût de rance" is perceived in from 10 to 15 seconds, and lasts for 50 or 60 seconds. The bitterness of some wines makes itself felt in 4 or 5 seconds, and persists for as much as 280 seconds.

In tasting, it should be kept in mind that certain qualities are liable to variations, according to the condition and age of the wine. The delicacy of a wine, for example, is almost totally hidden when the wine is young; the more so the younger the wine. This is due to certain substances which are proper to new wines, but which, later, are deposited and disappear from the composition of the wine.

Aromas are more or less intense, according to their origin and to the very variable circumstances under which they are formed.

The sense of taste is the final judge, and from its sentence there is no appeal. But how much careful consideration should be used before this judgment is pronounced; what a multitude of sensations must be considered, on all of which this judgment must be based!

The tongue, the cheeks, the gums, the anterior and posterior palates, the larynx, the nasal cavities, and to a certain extent the stomach, all contribute their separate sensations, which must all be taken into account. Besides these, the taster has also the sensations received by the eye and the nose. With all this varied testimony to consider, he should reflect deeply before delivering his verdict. For this reason, the taster, during the tasting and the few moments following, truly solemn moments, should be completely undisturbed by noise or otherwise.

A taster can sometimes conveniently express his verdict of quality by means of numbers; usually those from 1 to 10 are used, and correspond to the following expressions:

10—Perfect.
9—Almost perfect.
8—Quite good.
7—Relatively good.
6—Fair; sound, but not harmonious.

From 5 to 0 indicate various defects, according to their gravity.

## III.

### QUALITIES AND DEFECTS OF WINES.

The art of wine tasting, like every art or science, has a language of its own, without which the taster could not properly express his criticisms, nor compare his opinions with those of other tasters regarding the same wine.

This renders it necessary to define or explain the various terms that have been adopted by tasters to express the sensations experienced by their senses of sight, smell, and taste, during the examination of a wine.

FOAM (*Spuma*, It.; *Mousse*, Fr.).—When a wine is poured from one vessel to another, or agitated in any way, there forms a more or less abundant foam; that is, at the surface of the wine there are formed in greater or less quantities collections of little gaseous bubbles.

FINE FOAM (*Spuma di grana fine*, It.; *Mousse à perles fines*, Fr.).—The foam due to the formation of very small bubbles.

COARSE FOAM (*Spuma di grana grossa*, It.; *Mousse à grosses perles*, Fr.).—When the bubbles are larger.

EVANESCENT FOAM (*Spuma evanescente*, It.; *Mousse évanouissante*, Fr.).—Said of that which disappears immediately, or almost as soon as formed. As the old saying has it: "*Vino che brucia la spuma*" (a wine that consumes its foam).

PERSISTENT FOAM (*Spuma persistente*, It.; *Mousse persistante*, Fr.).— When the foam lasts some time and disappears slowly.

Persistent foam, as a rule, is characteristic of a wine poor in alcohol; of a wine at a low temperature, or of a wine in need of racking, or, it may be, of a wine which is undergoing a slow fermentation, which may be either the normal and necessary alcoholic fermentation, or may be what is known as a secondary fermentation, in which case the wine is a prey to some malady—tartaric fermentation, for example.

The foam may also be persistent on account of effervescence, that is, the continued giving off of carbonic acid, which is dissolved in the wine, and which in escaping on the decrease of pressure forms little bubbles which renew the foam.

In the first cases cited above, the foam is usually limited to a more or less imperfect crown or ring of bubbles which form around the edge of the glass; or if the wine contains more than the usual amount of carbonic acid a bubble of gas will now and then be formed and rise to the surface.

When some disease is the cause of the persistent foam, especially if it be that known as "*subbollimento, cercone, or vino girato*" (*vin tourné* of the French), the circle formed is called "*unghia*" (nail), from which the expression "*il vino fa l'unghia nel bicchiere.*" [This disease of *turned* wine is due to the filiform ferment, which destroys the tartar of the wine.—*Trans.*]

In the last case, when the persistent foam is due to effervescence, which may be of various intensities, several distinctions are made, of which the following are the principal:

SHARP, PUNGENT (*Frizzante, Piccante, Wine which has the Pinzo*, It.; *Mordant, Piquant*, Fr.).—In this case there is a somewhat abundant giving off of bubbles of carbonic acid when the wine is agitated, and even after, which tend to cling to the sides of the glass. Some one has written of a wine of this kind that "*nel berlo baciá e mordé*" (it kisses and bites); it makes itself felt as a smarting or pricking on the palate.

*"Sarà forse più frizzante
Più razente e più piccante.*"—Redi.

This pricking is caused by the presence of a larger amount of carbonic acid than is normal to the temperature and pressure.

The Tuscan usage of "governo" imparts this character to a wine.

"When the violent fermentation is over, throw in two handfuls of dried grapes to each vat; this will make the wine clearer and more piquant."—Davanzati.

However, Polacci rightly says: "For us a wine *governato* is always a defective wine."

FOAMING (*Spumeggiante*, It.; *Écumant*, Fr.).—This is said of wines which contain so much gas that when they are agitated bubbles are given off copiously, enough to form a layer of foam over the whole surface of the liquid. In the words of Redi:

*"Che nei vetri zampilla,
Salta, spumeggia, e brilla.*"

Wines which are bottled young, before they are well defecated, or which contain sugar when bottled, easily become "*spumeggiante*" when kept in a cool place.

SPARKLING (*Spumante, Mussante*, It.; *Mousseux*, Fr.).—This is said of wines which, after pouring into a glass, give off from every part an abundant supply of bubbles of carbonic acid, or foam, which collects at the surface and is continuously renewed for some time. The wine bubbles, and as is commonly said, pearls the surface.

In sparkling wines, the carbonic acid is in solution at a relatively high pressure.

In these wines, after the first violent ebullition of gas, there is what is known as the "*fontanella*," sparkling, which is due to a continuous development of very small bubbles of gas, which, starting from certain points at the sides or bottom of the glass, rise like little chains of beads to the surface, where they cause the phenomenon known as pearling.

Of sparkling wines there are three grades, based upon the amount of foaming, or rather on the amount of carbonic acid which is given off, and on the length of time during which the foaming continues.* These grades are:

CREAMING, GENTLY SPARKLING (*Mezzo spumante*, It.; *Crèmant*, Fr.).—These are wines in which only a slight layer of foam forms, and which give off but a moderate amount of gas; that is, effervesce very slightly.

The pressure exerted by these wines on the interior of the bottles is less than three atmospheres.

ORDINARY SPARKLING, OR MEDIUM PRESSURE (*Spumante, bella spuma*, It.; *Mousseux ordinaires*, Fr.).—In these wines there is sufficient gas to cause the foam to flow from the bottle the moment it is uncorked. The

---

* The French have a fourth grade, which they call Tissane, and which includes second and third-rate wines, which are, however, fairly drinkable.

pressure in this case varies from three to three and one half atmospheres.

STRONGLY EFFERVESCENT (*Molto spumante, Spuma forte*, It.; *Grand mousseux*, Fr.).—In these the cork is forcibly ejected from the bottle when unwired, and the wine is sufficiently charged with gas to be expelled from the bottle by its own pressure.

In these wines the pressure approaches or surpasses four atmospheres. The maximum pressure that bottles will stand, without great danger, is about six atmospheres.*

Either too low or too high a pressure is a serious defect in sparkling wines. If the pressure is too low they do not effervesce; if, on the other hand, the pressure is too great, as in the case of bottles which the French call " recouleuses," there is a ruinous percentage of broken bottles, or if the bottles do not burst the cork is driven out, and most of the wine lost.

The carbonic acid which is dissolved in these wines, is produced by the fermentation of added sugar, or of a portion of that which the must contained.

As already stated, wines which have been fermented dry, and not with a view of making them sparkling, can be rendered so afterwards by being charged, at a high pressure and low temperature, with carbonic acid. On this is based the system of Carpené, a system now much used both in Italy and abroad.

Sparkling wines may be:

SWEET (*Dolci*, It.; *Doux*, Fr.).—When the sweetness is decided and due to a large addition of syrup.

DRY (very slightly sweet) (*Semidolci, Dolcigni*, It.; *Douceâtres*, Fr.).— When the sweetness is slight or hardly noticeable.

EXTRA DRY (*Secchi, Asciutti*, It.; *Secs*, Fr.).—Which the English taste calls for; when there is no trace of a taste of sweetness.

In various red wines the foam may present different colors, as:

WHITE (*Bianca*, It.; *Blanche*, Fr.).—The case usually with old wines. There are, also, in some localities, young red wines of which the foam is white or whitish.

ROSE (*Rosea*, It.; *Rosée*, Fr.).—This is the case with lightly colored young wines, and is characteristic, it may be said, of mature wines.

RED, RUBY (*Rossa, Rossa rubino, Vermiglia*, It.; *Rouge, Vermeille*, Fr.). The color of the foam of heavy-bodied, deeply colored young wines.

ORANGE RED (*Rossa granato*, It.; *Rouge grenat*, Fr.).—This is a deep vinous red, resembling the color of pomegranates, and is often seen in cutting wines, or those blended with them.

BLUISH (*Turchiniccia, Bleuastra*, It.; *Bleuâtre*, Fr.).—Seen in wines poor in acid; as in some cutting wines which possess only from 3 to 4 per cent in acid.

BRIGHT, CLEAR (*Viva, Brillante, Smagliante*, It.; *Vive, Brillante*, Fr.). When the foam has a clear, crystalline appearance; this is generally seen in generous, young wines of full acidity.

DULL, DEAD (*Poco viva, Morta*, It.; *Morte*, Fr.).—The opposite of the

---

*As a rule, authors give higher figures for the pressure of the various kinds of champagne than I have indicated, but the fact is, that my figures, if not too low, are certainly not too high. Of this, I am assured by Professor Carpené, who, in his experiments with sparkling wines, had occasion to test the pressures of many wines from the best accredited foreign and domestic houses.

foregoing; indicates a diseased or decrepit wine, or one in need of racking.

After the foam is disposed of, the taster remarks on the degree of limpidity which the wine presents; a wine is said to be:

CLEAR (*Limpido*, It.; *Limpide*, Fr.).—When it is transparent and without cloudiness; or what Columella calls "*vinum defaecatum quam limpidissimum.*"

BRIGHT, BRILLIANT (*Brillante, Diafano, Lucido, Smagliante,* It.; *Brillant, Lucide, Luisant,* Fr.).—These terms are used to express a perfect and, as it were, crystalline transparency. This is the condition of wines that have been well clarified or filtered.

It may be noted here that clarification, unlike filtration, slightly modifies the composition of wine, as is proved by the quantitative determination of Professor Carpené, relative to wines that had been treated with white of egg. Following are the results of these determinations:

| | Tannin. | Œnocyanin. | Extractive Subtances. | Ash. |
|---|---|---|---|---|
| Wine of 1873, unclarified_____ | 0.91 | 0.42 | 21.39 | 3.12 |
| Wine of 1873, clarified_____ | 0.41 | 0.24 | 19.91 | 3.06 |
| Wine of 1874, unclarified_____ | 1.15 | 0.82 | 24.22 | 2.80 |
| Wine of 1874, clarified_____ | 0.57 | 0.44 | 20.17 | 2.79 |

The quantity of albumen employed was about 100 c.c. per hectolitre (1 per m., or 1 pint to 125 gallons), which is a usual dose.

CLOUDY, DULL (*Vellato, Appannato,* It.; *Voilé,* Fr.).—This is said of wines that are not quite clear, that show a slight cloud or dimness, due to the presence in them of substances in suspense in a very fine state of subdivision. This is noticed, for example, in wines recently racked, especially when, during the operation, they have been much exposed to the air and drawn into well-sulphured barrels.

This slight defect, which is easily cured, is also frequently found in wines made from grapes grown on rich soil, and also in wines which, being poor in acid, have not undergone a complete fermentation.

Wines, of course, may possess different degrees of cloudiness, which are generally expressed by the terms cloudy, slightly cloudy, nearly clear, etc.

TURBID, MURKY, THICK (*Torbido,* It.; *Trouble, Cassé,* Fr.).—When the suspended particles are large enough to be almost visible to the naked eye, and present in sufficient quantity to completely destroy the transparency of the wine and make it almost opaque.*

---

* Old bottled wines may be turbid either because they have become unsound, as happens very easily when bottled too young, or because they have not been thoroughly defecated before being bottled, or it may be, because they have been moved in such a manner as to stir up the slight deposit which all wines throw down in time in greater or less quantities. If the wine is unsound there is no need of precautions, for the wine has become undrinkable; if, on the contrary, the turbid wine is sound it must be moved with the greatest caution, and to prepare it for the table it will be found useful to follow the rules of C. Ladrey, who writes thus:

"When the time arrives to drink a wine which has lain in bottle for some years, the first thing to do is to examine the bottle with great care when it is lifted up. It should be lifted up cautiously, retaining it in its horizontal position. By carrying the wine into the light, daylight or artificial, it is easy to ascertain whether the wine is perfectly clear or has a deposit. If, as may happen, the wine be perfectly clear, without trace of deposit, the bottle may be stood up and the wine served from it without decantation. This case, however, is very rare, and, especially with old wines, there is generally a deposit.

A wine from low-land grapes, in which tartaric fermentation has reached the stage of development when carbonic acid begins to be freely given off, is a good example of this condition.

This defect may be simply transitory, as when a wine has lately received some treatment, or an addition of alcohol or tartaric acid, or directly after cutting or mixing wines, or when a wine has been much shaken or been exposed to too low a temperature. If the defect is permanent, it shows that the wine is diseased or ready to become so, or that the wine has been badly made. In the former cases the wine simply needs time to depose or an increase of temperature, when it will right itself. In the latter cases some special treatment is necessary, such as sulphuring, addition of tartaric acid, clarification, pasteurizing, etc.

OPALESCENT, IRIDESCENT (*Cangiante, Opalescente, Iridiscente,* It.; *Chatoyant,* Fr.).—When the light in passing through the wine is decomposed, that is, when in looking through the wine rays of different colors are seen. This iridescence is best seen at the surface of the liquid and near where it is in contact with the glass; it is due, not to reflection or refraction, but to the phenomenon of interference.

A wine exhibiting this peculiarity is open to grave suspicion of unsoundness, if it is not already in an advanced stage of disease.

As an example of a wine in this condition, may be cited one which is, in the first phases of the disease, known as "subbollimento."* If a little of this wine is left exposed to the air it first becomes turbid, and loses its red color; then a precipitate forms and leaves a yellowish, sour, somewhat bitter liquid on top. As the disease progresses, if the wine is slightly shaken, mucous clouds will be seen floating in it, at the surfaces of which the above-mentioned phenomenon of interference may be seen.

In the time of Pliny, to describe the color of a wine they had only the four following epithets: *album, fulvum, sanguineum, nigrum.*

In those days they were easily satisfied; now we use the following terms to describe the colors of red and white wines:

COLORLESS, DECOLORIZED (*Incolore, Scolorito, Decolorato,* It.; *Incolore, Decoloré,* Fr.).—When the wine has almost the appearance of pure water; when the rays of light pass through it without suffering any or only imperceptible changes.

Colorless wines are easily obtained from perfectly ripe white grapes, picked and handled with great care, and crushed when quite fresh and quite cool; then by exercising the most scrupulous cleanliness during the vinification and keeping of the wine, and by fermenting the must after it has been well defecated. If a wine is made which is not per-

---

In this case we must be careful not to mix the limpid part of the wine with the deposit, and before raising the bottle up the wine should be decanted, which in its result is an operation exactly similar to racking. This decantation should be made in the cellar, and demands some precautions. First the neck of the bottle is carefully raised, but not too high; it is then uncorked, care being taken not to subject it to any brusque motion either in raising it or in drawing the cork. The wine is then poured into another perfectly clean bottle, taking care to stop before the smallest part of the deposit has passed into the fresh bottle or decanter. The quantity of wine lost by this method is very small, and the wine that is saved can be drunk to the last drop. If, on the contrary, a wine which has only a very slight deposit is placed on the table without decanting, the second or third glass will commence to show a loss of brightness and the wine will have lost its agreeableness. There are some very simple machines made, which work on the principle of the siphon, and which greatly facilitate the operation of decantation."

*"La pousse" of the French, a kind of tartaric fermentation which is fully described on a subsequent page.—*Trans.*

fectly colorless, it may be rendered so by the use of animal charcoal, properly prepared, that is to say, in such a way as to prevent its diminishing the acidity of the wine. If this precaution is not taken, the wine, on account of its diminished acidity, will quickly turn yellowish on account of the formation of ferric compounds, which, under these conditions, takes place with great readiness.

STRAW-COLORED (*Paglierino,* It.; *Couleur de paille,* Fr.).—Of the color of straw, but somewhat pale.

AMBER, YELLOW (*Giallo,* It.; *Jaune,* Fr.).—Is said of wines which have a deeper straw color.*

GOLDEN, GOLDEN-YELLOW (*Giallo dorato, Aurato, Dorato,* It.; *Doré,* Fr.).—This epithet sufficiently explains itself.

"*Egli è il vero oro potabile,*" wrote Redi of the wine of Trebbiano.

GREENISH (*Verdognolo, Verdiccio,* It.; *Verdâtre,* Fr.).—When a wine has a slight greenish tint, resembling somewhat the green of grass. This color is characteristic of certain varieties of grapes; for example, the Verdea or Bergo.

Regarding the wine of this variety, it is said that the Verdea of Tuscany is not so called on account of its green taste, but because of its greenish tint.

Frederick the Great, of Prussia, had a great predilection for the wine of Verdea.

This greenish color is also characteristic of the wines of Reno, and in general of wines made from somewhat acid grapes.

PINKISH-YELLOW, OR PINKISH STRAW-COLOR (*Paglierino rossastro, Giallo rossastro,* It.; *Paille roussâtre,* Fr.).—Sometimes a wine, in addition to

---

* This yellow color may be natural and proper to the wine, or it may be a color which it has acquired from several causes, among which are some that have very grievous effects on the wine, and may be considered properly as maladies.

The wines most generally subject to this disease of becoming yellow are those poor in alcohol, tartar, tartaric acid, and tannin, and which on the other hand are rich in malic acid.

I have already alluded to one of these causes above, namely, the presence of iron compounds. Some colorless wines, which are rather poor in acid, become, when placed in contact with the air, yellow or yellowish brown, in consequence of the formation of complex compounds, ferric, humic, etc.

The commonest causes of the yellowing of wines can be traced to the conditions under which the vintage has taken place; if, for instance, the season has been cold and rainy, and the grapes have been gathered after the vines have in great part been denuded of their foliage, if the bunches contain decayed, soft, insipid grapes poor in acid and sugar, a wine of poor keeping qualities is obtained, and one very likely to become yellow, unless art comes to the aid of nature.

Robinet, who has made special investigations with regard to the causes of this deterioration of white wines, distinguishes between that due to a fermentation caused by a mycoderm, and those due to chemical action, and among the latter he mentions some which give rise to the formation of malic ether, which reacts on the sugar. I should, however, remark here that after stating his belief in the formation of the malic ether, he declares that he has been unable to find the rational equation of the reaction, or definite proof of its existence, but bases his belief in the formation of the malic ether on the taste and pronounced odor of cider which the wine acquires—an odor which is characteristic of the above substance.

Robinet also makes the important observation that during his researches he had noticed the disappearance of the glycerine from wines which were becoming yellow. This disappearance of the glycerine would lead one to believe that the reactions which take place are much more complicated than supposed by Robinet, especially in consideration of the fact that the glycerine is subject to transformations, like the other ingredients of wine.

Instead of trying to cure or ameliorate this defect in wines, it should be prevented, which can be done by the addition of alcohol and acids.

The secondary fermentation which causes this disease is due, still according to Robinet, to a particular mycoderm, which can be seen distinctly with a magnifying power of nine hundred diameters. This mycoderm is extremely small, and of an oblong shape; it is $\frac{1}{500}$ m.m. in length, and $\frac{1}{800}$ m.m. in width.

its yellow or straw color, will have a pinkish tint of more or less intensity. This may be considered as due to imperfect cleanliness of the vessels used in wine making, or of the barrels in which the wine has been put.

ROSE-COLORED, SHILLER (*Rosato,* It.; *Rosé,* Fr.).—White wines made from red grapes frequently possess this color in greater or less degree; especially is this the case when the grapes have not been picked and handled with great care, or when the grapes have become the least heated.

A white wine may also acquire this color by contact with barrels or utensils which have been used for red wine and not been thoroughly cleansed afterwards.

This color is sometimes produced artificially. In France they use extensively *teinte de Fismes,* so called after the town in which it is manufactured. It is claimed that it is free from alum and sulphuric acid,* but wrongly.

White wines which have commenced to spoil, or in which viscous fermentation has started, and which begin to become brownish, or even bluish, and at the same time turbid, what the French call *vin oeil de perdrix,* are rendered salable by the use of this *teinte de Fismes,* and are sold by the French under the name of *vins rosés.*

Jacquesson, *père,* states that this coloring fluid not only colors and clarifies the wine, but also arrests the progress of the disease, or prevents it if it is to be feared. This fluid is also used in France for coloring sparkling wines.

BLUISH-BROWN, BROWN-YELLOW (*Bruno-bleuastro, Giallo-bruno,* It.; *Brun-bleuâtre,* Fr.).—This color, which the French call *oeil de perdrix* (partridge-eye), is a dull, dark yellow, proper to some old, southern wines, but due in the majority of cases in which it is found to some malady of the wine. †

This phenomenon is observed not only in old but also in young wines, both red and white. Very probably its origin lies in several causes, as the numerous explications given by different authors would lead us to believe. Nessler has studied the change of color as it takes place in white wines. He tells us that the substances that cause the coloration, more or less deep, of the wine are contained in the stems and the seeds. Thus, wines which have been fermented in contact with the solid part of the grapes blacken very easily when exposed to the air. The presence of bad grapes in the fermentation also tends to render a wine liable to this discoloration.

This change of white wine depends directly on the action of the air;

---

* The *vin,* or *teinte de Fismes,* was first prepared by Manceau by boiling elderberries and cream of tartar together.

† It sometimes happens, writes Robinet, that a perfectly bright white wine which has never been racked or otherwise treated before, is racked from its lees and treated with tannin and some clarifying material; then instead of becoming bright and clear the operations to which it has been treated have had diametrically the opposite effect. The wine has not taken the clarification, as the cellarmen say, has a bluish tint, and is turbid.

This change or malady of the "blue color" happens most generally in wines of low acid and alcoholic contents, and which are at the same time rich in nitrogenous substances. According to Robinet this malady is due to a secondary fermentation, caused by a mycoderm which is analogous to the *mycoderma crocceum,* and has a very ephemeral existence.

To cure this disease in a wine it generally suffices to raise the alcoholic strength, or sometimes an addition of six or eight grains of tannin per hectolitre is necessary. In the latter case the wine is allowed to settle for twenty-four hours after the addition of tannin, and then clarified with isinglass.

The above mycoderm is killed and precipitated by cold.

the wine loses its limpidity, becomes cloudy, and a black precipitate is formed; meanwhile the taste of the wine often changes.  The black substance may be decolorized by sulphurous acid; the use of this substance arrests or retards the blackening of the wine.

Wines made from grapes poor in tartaric, malic acid, etc., like those which have been gathered when wet with dew or rain, or those which have been injured by cryptogams, are liable, when exposed to the air, to become cloudy and dark in color.

The presence of an excess of iron in the white wines of certain localities of the southern provinces is the reason why, when they are at all exposed to the air, their color changes to a blackish green.

Not southern wines alone, but also those from northern provinces, when they do not contain a sufficient quantity of acid, and more especially of tartaric acid, acquire this color.  Chemists explain this phenomenon in different ways, though all admit that it is due to the presence of some of the compounds of iron.  Nessler tells us that wines produced on soils rich in the salts of iron, and even wines which have been for any length of time in contact with iron, as happens when there is an iron rod between the heads of the cask, or when there are nails in the cask, etc., if they become exposed to the air, turn black, for then the protoxide or ferrous oxide contained in the wine changes in contact with the air to sesquioxide or ferric oxide.  A black compound is then formed by the combination of the ferric oxide with the tannin; this black color is not obtained with the protoxide.  Other chemists explain the phenomenon by supposing that there occur or are formed in the wine certain humic products analogous to those which are formed by the decomposition of vegetable substances.  These substances are feebly acid, and have a considerable dissolving power on the iron.  Thus there are formed in the wine certain of the lower compounds of iron, which, on exposure to the air, change to the higher compounds, and give the wine the blackish tint before spoken of.  The wine then becomes turbid, and the flavor undergoes certain peculiar changes.

Formerly some sparkling wines were made of this color, but now it is no longer found but as a defect.

DIRTY (*Sporco*, It.; *Terne*, Fr.).—A diseased, badly made, or badly kept wine sometimes becomes turbid, and its natural color is masked by other colors, giving the impression of something soiled or dirty.

Among red wines the following are the colors most generally recognized; they may be of more or less intensity:

VERY LIGHT RED (*Claretto. Chiarello, Chiaretto*, It.; *Clairet*, Fr.).— These terms are used to describe a class of wines which contain the least color of any red wines; the cause of this poverty of color may be in the nature of the grape, the mode of preparation, or it may be that the wine has been diluted with water.

These wines form the connecting link between white and red wines.

Trinci, writing of these wines, says: "The French 'claretto' is a smooth, vinous, lightly colored wine, with little aroma; slow and long in maturing, and not pleasing when drunk alone; blended, however, in proper proportion, it is extremely good."

The "claretto" drunk by Redi, however, must have been very different from this, or he would not have written:

> " *Benedetto*
> *Quel claretto*
> *Che si spilla in Avignone.*"

3

RUBY (*Rubino*, It.; *Rubis*, Fr.).—Wines which have a fine, vinous red, which recalls the color of the ruby.

This color is that found most commonly in table wines; for instance, the wine of Chianti; it is also the color of the wines of Bordeaux.

Some writers speak of vermilion wines, but a wine is never really of that tint; wines rich in acid and of bright, intense ruby, will appear for the moment to be vermilion immediately after being racked, on account of the presence of a slight cloudiness.

PURPLE (*Porporino*, It.; *Pourpré*, Fr.).—The case where the natural red of wine tends slightly to violet.

This color is seen in Montepulciano when it has reached perfection.

GARNET, RED (*Granato, Rosso cupo*, It.; *Rouge sombre*, Fr.).—Said of wines which have a more or less intense blood-red, recalling the color of garnets and similar precious stones, and of some varieties of gooseberries, etc.

This garnet tint is seen in heavy-bodied dinner wines, such as Barbera, Gattinara, Borgogna, and in wines made from grapes grown on clayey and ferruginous soils. These wines in aging are apt to acquire more or less of the orange tint.

BLACK (*Nero*, It.; *Noir*, Fr.).—This color, the *nigrum* of the Romans, is really never found in wine; the darkest wines, made from the Teinturiers, are not quite black, nor is even the concentrated solution of œnocyanin obtained by the Carpené-Comboni process.

VIOLET, BLUISH (*Violaceo, Turchiniccio, Bleauastro*, It.; *Bleauâtre, Violacé*, Fr.).—This color is seen in a more or less marked degree in blending and other wines poor in acid. This tint is due to the violet coloring matter which is contained in certain dark wines of southern Italy. It is very unstable, and precipitates with great readiness. It is also found in the wines from certain American coloring grapes, such as the Jacquez, the Marion, and York's Madeira, when they have been made without addition of plaster or tartaric acid.

ORANGE, YELLOWISH-RED, RUSTY (*Aranciato, Giallo aranciato, Color matone, Rossico*, It.; *Orangé, Pelure d'oignon*, Fr.).—These are the colors or tints of old or decrepit wines. By decrepit wines should be understood wines which have passed their prime and have begun to lose their valuable qualities.

These tints are seen sometimes in young wines, but less marked than in old; especially in those which, at first, have much of the bluish tint, and which deposit their color quickly.

Old wines often lose all, or nearly all, of their color, and become what is called "scolorito," decolorized or faded.

DARK COLORED (*Colorato*, It.; *Coloré*, Fr.).—Said of wines that have relatively a great deal of color.

Wines may be divided according to intensity of color into deep-colored, medium-colored, and light-colored wines.

Deep-colored wines are harsh and indigestible.

I will now pass in review the qualities and defects of which the senses of taste and smell take cognizance.

AROMA (*Aroma*, It.; *Arôme*, Fr.).—By aroma must not be understood simply those odors which are delicate and agreeable, as when speaking of bouquet; for example, the foxy odor or aroma of certain American grapes, varieties of the species *Vitis labrusca*, and of the wine made from them, is far from agreeable.

The aroma is the odor which comes from the skins of aromatic grapes,* and varies in quantity and quality, according to the variety of grape and the degree of its maturity. It passes into the wine in wine making; the aroma therefore exists in the grapes as well as in the wine.

BOUQUET† (*Profumo*, It.; *Bouquet*, Fr.).—Every fine wine exhales an odor peculiar to itself, which is always delicate and pleasing. Exception may be made of artificial bouquets, which, if not absolutely disagreeable in themselves, are always too strong and intense in a wine.

The bouquet is due to the volatilization at ordinary temperature of certain substances known as ethers, which are formed by the reactions of the acids and alcohols in the wine during its process of aging.‡

Thus, the bouquet is not to be found ready formed in the grape, as is the case of the aroma.

SÈVE (*Abboccato*, It.; *Sève*, Fr.; *Göhr*, Ger.).—The "sève" is neither bouquet nor aroma; it is a certain savor, a certain fragrant quality of the wine due to a smooth and delicate blending of perfections, of aromas and bouquets, which is perceived when the wine is in the mouth and in the act of swallowing, affecting the olfactory organs through the internal nasal ducts. The bouquet and aroma affect the senses before, the sève after drinking the wine.

Carpené, writing of Moscato de Segesta, says: "Of the most delicate fragrance and exquisite flavor. It is a dainty, fruity wine, which fills the mouth with an harmonious ensemble of delicious flavors, which cannot be described, but can only be experienced."

Sève, which is especially the property of fine wines, is due to the presence of certain substances which are formed in the grapes during

---

* The ancients held aromatic wines in high estimation. They added to the must, during fermentation, different varieties of apples, then cane, amomum, cassia, saffron, ginger, and other species of aromas, to communicate the odor that they desired.

The aroma most highly appreciated was that obtained by the addition of myrrh. We read, in fact, in Pliny: *Lautissima apud priscos vina erant myrrhæ odore condita, ut adparet Plauti fabula, quae Persa iscribitur, quamquam in ea et calamos addi jubet.*

Peppered wine, which was prepared by fermenting the must with apples and pepper, was very much appreciated in the time of Pliny.

† Even the bouquet of wines has not escaped imitation and adulteration. The manufacture of artificial bouquets or perfumes for wines has become a regular industry in France and Germany, where it is carried on on a large scale. There is a large consumption of such articles as "bouquet" of Pomard, or of Bourgogne, extract of Bourdeaux, the "Rancio des vins," "sève" of Baumé, of Médoc, of St. Julien, of Champagne, of Sillery, etc.

The substances most usually employed to add an artificial bouquet to dinner wines, are: Florentine iris, raspberries, cloves, vine flowers, mignonette, nutmegs, bitter almonds, etc. To these should be added certain chemical products which are prepared more especially in Germany. All these attempts to imitate nature have been but very partially successful.

A wine may be perfumed artificially, but it is impossible to give it "sève." This artificial perfume is always too pronounced, and is never as delicate as the natural bouquet of wine. These artificial bouquets impress the sense of smell, but not that of taste. If a perfumed wine, then, is tasted without being smelled, its natural "sève" can be distinguished. Artificial aromas are not lasting, and gradually disappear from the wine.

‡ Chemically, the difference between aroma and bouquet is, according to Maumené and Berthelot, the following:

The former is due to certain hydro-carbons and to the products of their oxidation; perhaps, also, as Ordonneau states, to the ether of a high, fatty acid produced by inter-cellular alcoholic fermentation, and which, being fixed, remains in the pellicle; this has enabled the experimenter to obtain it from the pomace of Folle Blanche.

The latter seems to be due to a mixture of aldehydes with one or more essential oils and of numerous ethers, the product of the combination of fatty and other polyatomic acids with ethylic and other alcohols; there are, for instance, valerian, amylic, propyl-acetic, etc.

the short time preceding their complete maturity; these substances are peculiar to certain varieties of grapes, and owe their existence also to careful cultivation, as well as to certain conditions of climate and soil.*

In analyzing wine, writes Fauré, I have observed that fine and delicate wines, those renowned for their flavor and general high quality, contain a certain glutinous, viscid substance, which exists only in almost inappreciable quantities in ordinary wines, and is quite absent from inferior ones.

This principle, to which wine owes its sève, has been called by Fauré œnanthin,† or flower of wine, and is only found in grapes which are completely mature. Some vineyards, which usually produce grapes containing this substance, fail to do so in stormy seasons. The only vines containing it in such years are those produced on dry sandy or gravelly soils. The same variety of vines, which, when grown on an appropriate soil, gives a wine full of sève, will, when grown on a rich, heavy, or clayey one, produce a wine containing little or no œnanthin.

Thus it can be seen that the preëminence of high-class wines is not due to the caprice of the taster, but to actual differences of composition, and to the presence of principles not found in inferior wines.

The ordinary wines of the three communes of the Gironde, where the four high-class Bordeaux wines are produced, are, in general, poor in œnanthin. These four wines, however, contain a larger quantity of the substance, as may be seen by the following:

Œnanthin contained in vines of—

| High Class. | | Ordinary. | |
|---|---|---|---|
| Chateaux Margaux | 1.25 | Margaux | 0.70 |
| Chateau Lafite | 1.20 | Pauillac | 0.75 |
| Chateau Latour | 1.10 | Pessac | 0.50 |
| Haut-Brion | .65 | | |

FLAVOR (*Sapore*, It.; *Saveur*, Fr.).—In this character we have the effect of the wine on the sense of taste, and more particularly on the tongue, which best distinguishes between various tastes. The flavor is distinct from either aroma, bouquet, or "sève"; unlike the last, it does not affect the sense of smell. As has been shown, the sève is perceived after the wine has passed the base of the tongue, the soft palate; the taste, on the contrary, or better, the flavors, are perceived almost immediately, and continue to affect the tongue and its sides, or posterior

---

* The result of many observations and studies regarding the influence of soil composition or the character of wine, may be summed up as follows: High alcoholic strength is characteristic of wines grown on calcareous soils; color depends on the iron in the soil; smoothness on the alumina and on the variety of grape; bouquet on the silica.

Chambertin, writes Julian, is a wine which has a good color, much sève, is very delicate and smooth, faultless in taste, and possessing the most agreeable bouquet. The vineyard which produces this wine has the following soil composition:

| | |
|---|---|
| Alkaline salts | 0.031 |
| Carbonate of calcium and magnesium | 4.425 |
| Ferric oxide | 2.961 |
| Phosphoric acid | 0.235 |
| Alumina | 2.063 |
| Silica (soluble) | 0.110 |
| Organic matter | 1.973 |
| Insoluble residue (silica) | 89.302 |

† By œnanthin should not be understood, as perhaps was done by Fauré, a single chemical compound, but rather a complex mixture of ethers.

portion, with a series of sensations which are agreeable or disagreeable, according to the nature of the flavors and their degree of intensity.*

NEUTRAL FLAVOR (*Sapore neutro*, It.; *Saveur neutre*, Fr.).—A wine is said to be neutral when it has no marked aroma or taste.†

Wines of neutral taste are the best base for the making of imitative wines, as they acquire most easily the taste of the wines with which they are blended.

VAPID, FLAT, INSIPID (*Insipido*, It.; *Plat*, Fr.).—A wine is vapid when it is lacking in alcohol and vinosity, or when, without having any defect

---

* With regard to tastes in general, writers are at variance. The greater or less number of tastes and the possibility of their classification have been discussed.

The number of tastes may be considered as infinite, and therefore a classification almost impossible. Such classification, however, has been attempted. Haller distinguishes twelve tastes, which have been reduced by Linnæus to ten: sweet, acrid, fatty, astringent, bitter, viscous, saltish, watery, and insipid.

Vintschgau proposes another taste—metallic.

Physiologists distinguish in the sense of taste four specific energies, that is, four elementary sensations, viz.: sweet, bitter, acid, and salt. The first two affect only the nerves of taste; the acid taste, on the other hand, if too strong, may cause pain, for which reason Vintschgau believed that acid and salt tastes affect also the sense of feeling, as is seen in touching concentrated solutions of acids.

Nothing is known with certainty as to the way in which different tastes are distinguished, and we must be content with supposing that each flavor—sweet, sour, bitter, salt—acts upon special nerves which serve to distinguish them. This is the more probable, as different parts of the tongue are unequally affected by different tastes. We are still more in the dark regarding the intimate nature of the tastes, the chemical composition of the substances which they characterize seeming to have no connection with them.

The chemical composition of a substance has nothing to do with its sweet, bitter, or salt taste; with regard to the acid taste, however, it may be said that every substance which tastes acid is also an acid from the chemical point of view.

† The vineyardist in making a choice of varieties to plant should keep in view the flavor which they will give to his wine. If he is planting in a new locality, where it cannot be known what kind of grape will there best develop its flavor, he should choose a variety which gives a wine of neutral taste.

The French, who are masters of the art of imitating wines, have this maxim: "There are more buyers than there are connoisseurs."

Trusting to the truth of this saying, they have been able to establish that great commerce of wine which has become one of the principal sources of riches to France. The cities of Cette, Bordeaux, Marseilles, Lunel, Montpellier, and others of the south of France are centers of the production of large quantities of "wines of imitation."

Do you wish to make, for example, a hectolitre of fine Bordeaux?

Take—

Red wine of the south (Roussillon or Narbonne) ................................60 litres.
White wine of good quality ................................................25 litres.
Old wine of Alicante ........................................................12 litres.
Old wine of Malaga............................................................ 3 litres.
"Conservatore enantico" ...............................................25 grammes.

The œnanthic conservative is dissolved in about a litre of warm white wine; the whole is then well mixed and allowed to stand for two weeks. During this time a slow, insensible fermentation goes on, which completely mixes or blends the ingredients.

The wine is then drawn into sulphured casks, clarified, racked again, and the Bordeaux is made.

This, however, is too expensive a Bordeaux; here is a cheaper one:

Red common Spanish wine ...................................................70 litres.
Wine of Narbonne ..........................................................25 litres.
Wine of Malaga ............................................................ 5 litres.
Bordeaux extract..........................................A quarter of a bottle.
Œnanthic conservative......................................................30 grammes.

This is treated in the same way as the first.

If a still cheaper Bordeaux is desired—

Ordinary red wine ..........................................................81 litres.
Roussillon and Narbonne.....................................................15 litres.
Old brandy ................................................................ 4 litres.
Bordeaux extract ..........................................A quarter of a bottle.
Œnanthic conservative.......................................................30 grammes.

The above information is for the edification of those who prefer a bottle of this Bordeaux to a bottle of Chianti, of Valpolicella, of Valtellina, and of many other Italian wines which are far superior to these French concoctions.

due to secondary fermentations, it lacks some of those qualities which together render a wine agreeable.

An insipid wine may have plenty of color, however. Insipid wines are very subject to unfavorable changes.

SAPID (*Sapido*, It.; *Sapide*, Fr.).—A wine is described as sapid; it is meant that the acids are agreeable in quality and proportionate in quantity.

VINOUS, VINOSITY* (*Vinoso, Vinosita*, It.; *Vineux, Vinosité*, Fr.).—A wine is said to possess vinosity when it imparts in a certain degree that sensation of warmth characteristic of the alcoholic flavor.

WEAK (*Debole, Vino che scappa in bocca*, It.; *Faible, peu alcoolique*, Fr.). A wine is said to be weak when it is of low alcoholic strength, or when its alcoholic contents are not in proportion to its other constituents. Wines of this character have in general little flavor, are insipid, and difficult to keep, on account of the gummy or mucilaginous substances which they contain, and to which they owe what little flavor they have.

LIGHT (*Leggero, Sottile*, It.; *Leger, Mince*, Fr.).—A light wine is one which is of good quality, but at the same time contains a relatively small amount of color, body, and alcohol, no prominent flavors, and no sweetness. The general effect of a light wine is one of delicacy, though there exists a just equilibrium between the various constituents.

SOFT, MILD (*Molle*, It.; *Mou*, Fr.).—A mild wine is one which does not affect the palate by its harshness or astringency, as do rougher wines. Softness characterizes wines which are neither sweet nor dry, and not too alcoholic.

ALCOHOLIC (*Alcoolico*, It.; *Alcoolique*, Fr.).—When a wine is spoken of as alcoholic, it is generally meant to be one containing a relatively high per cent of alcohol, but of an unsatisfactory and unsatisfying quality.

GENEROUS (*Generoso*, It.; *Genereux*, Fr.).—A generous wine is one with plenty of alcohol, but of a smooth, warming, strengthening character; one of which a small glass produces a feeling of well-being and sensible tonic effects.

WARM, HOT (*Caldo*, It.; *Chaud*, Fr.).—A hot wine is one containing a good deal of alcohol, which produces a somewhat burning sensation in the mouth and stomach.

SHARP, LIVELY (*Vivo*, It.; *Vif*, Fr.).—This is said of a wine which, without being pronouncedly acid or alcoholic, affects the palate vividly. It is a quality compatible with lightness, but not with smoothness.

FULLNESS, ROUNDNESS (*Stoffa*, It.; *Étoffe*, Fr.).—Expressive of a robust homogeneity, which gives the impression of solidity and good constitution.

BODY (*Corpo*, It.; *Corps*, Fr.).—A wine is heavy bodied when it is rich in extractive matter and has high vinosity.

HEADY (*Fumosa*, It.; *Fumeux*, Fr.).—Wines which contain much carbonic acid, and thus go quickly to the head, produce effects that are usually confounded with those of drunkenness, but which, in reality, differ very much from them physiologically. Wines of this character are unwholesome.

---

* Many use this word in a somewhat different sense; by it they mean "wine-like;" that is, having a full supply of the quality or qualities which preëminently distinguish wine from other alcoholic beverages.—*Trans.*

DENSE, PULPY (*Carnoso, Polputo, Maccherone*, It.; *Charnu, Pulpeux, Lourd*, Fr.).—Expressive of a wine that has what one might almost call a pasty consistency.

HEAVY, COARSE (*Grave, Gravone, Pesante, Capitoso*, It.; *Lourd, Gros, Pesant, Capiteux*, Fr.).—Wines which have much body and little alcohol, and which, even when drunk in small quantities, go to the head and weigh on the stomach.

CLEAN (*Franco*, It.; *Franc*, Fr.).—Said of a wine which does not leave the slightest suspicion of any taste indicating unsoundness, or of any defect due to the bad condition of the grapes from which it was made, or to neglect or improper handling of the wine.

HARMONIOUS (*Armonico*, It.; *Harmonique*, Fr.).—Well constituted. This is said of a wine when its constituents are in exactly the proper proportions, well balanced and blended, forming a perfect whole, which is at the same time pleasing and satisfactory.

WINE THAT ENDS WELL (*Vino che finisce bene*, It.; *Vin qui finit bien*, Fr.).—This is an expression used by the taster to define an impression that remains for a certain time after drinking a fine wine; it means a wine in which the constituents are harmonious, and remain so even after the wine has passed from the mouth, impressing the senses with nothing but pleasing sensations to the end. These sensations continue even after the wine has been swallowed, insomuch that one might almost say that it wished to prolong the pleasure of the drinker by a fresh visit to the organs of taste.

WINE THAT ENDS QUICKLY (*Vino che finisce presto*, It.; *Vin qui finit vite*, Fr.).—Wine that leaves but an ephemeral sensation in the mouth; that is to say, almost as soon as the wine is swallowed all trace of it is gone, and the palate, tongue, and stomach seek in vain to recall its character, flavor, bouquet—all have gone, all have disappeared.

WINE THAT ENDS BADLY (*Vino che finisce male*, It.; *Vin qui finit mal*, Fr.).—A wine that after swallowing leaves a disagreeable taste, bitter, woody, etc., in the mouth.

DELICATE (*Delicato*, It.; *Delicate*, Fr.).—A wine to be delicate must be perfectly harmonious, soft, and agreeable.

FINE, OR HIGH QUALITY (*Fino*, It.; *Fin*, Fr.).—A wine that unites a natural delicacy with an exceptionally agreeable flavor and delicious bouquet.

MUTE (*Muto*, It.; *Muet*, Fr.).—Said of unfermented or only partially fermented wines; they are characterized by a sweetish or gummy taste. They are wines which have been made from musts treated with sulphurous anhydride or fortified with alcohol. The wines that are generally made "mute" are white wines that are to be used to sweeten liquors or to increase the sugar contents of new wines, or that are to be used for the manufacture of syrups by concentration in vacuo.

When a wine is made mute by the use of sulphurous anhydride, the risk is run, if too much is used, of giving the wine, first, a taste of sulphydric acid, and afterward more or less pronounced bad flavors due to the sulphates that are formed.

These wines are kept in cool cellars, where the temperature is as nearly as possible constant, and in strong and well-hooped casks. They ought to be clarified, preferably with gelatine. In order to obtain a perfect clarification, about 8 or 10 grammes of tannin are added to each hectolitre before putting in the finings (one tenth per m., or about 1.25 ounces per 100 gallons).

SMOOTH (*Vellutato, Morbido,* It.; *Velouté, Moelleux,* Fr.).—A smooth wine fills the mouth with its grateful flavors and fagrance, imparting its delightful series of sensations without the slightest harshness.

This quality is due to the presence of a certain quantity of glycerine, and not to glucose, as at first one might be inclined to think. In this latter case the wine would be called "amabile" (fruity).

It is glycerine rather than glucose which gives a wine that kind of smoothness which might almost be called unctuosity.

In very high-class wines the smoothness or unctuosity is due not only to glycerine, but also to other bodies which have not yet been well studied; they occur more especially in wines of very favorable years; that is, of years when the season has been so propitious that the grapes have been able to attain an exceptionally perfect maturation.

Many chemists have attempted to determine the nature of these substances.

Il Fauré, who studied the wines of the Gironde, believes that this unctuosity is due to the same substance as sève, a substance which is of similar character to pectine and mucilage, and which he called "œnanthin."

Batilliat claims to have found in the high-class wines of Bordeaux the peculiar substance which causes their unctuosity, and which he calls "croatine."

Mülder, on the other hand, from observations made on the wines of the Gironde, considers this unctuous substance as analogous to dextrine.

Whatever may be the nature of this substance, it is useful to know that the wines in which it occurs, if not well kept, are liable to undergo an almost insensible fermentation, which destroys this substance, and so takes away from the wine that quality which is due to it; pasteurizing or heating will also deprive a wine of this quality.

FRUITY (*Amabile,* It.; *Suave,* Fr.; the Latin, *Suavis vel subdulcis*).— A wine which is very faintly sweet on account of retaining a small quantity of grape sugar or glucose.

As is said sometimes: "*Quel vinetto cosi amabile va giù senza accorgersene.*"

Technically, a fruity wine cannot be said to possess sève because it tends towards sweetness. However, a wine which is very slightly sweet may possess a good sève in the sense that it produces those sensations which are the quality of wines of the highest class.

SWEETISH (*Dolcigno,* It.; *Doucereux,* Fr.).—A wine is said to be sweetish when its sweetness is undecided, unsatisfactory, and not in harmony with the other components of the wine; it is due usually to a bad fermentation and incomplete defecation, or it may be, with an ordinary table wine rich in mucilaginous substances, that it is becoming sick or undergoing one of those insensible fermentations, that is, the tartaric fermentation, to which such wines are so subject in the spring. In the latter case there is a moment when the wine can be detected in becoming slightly sweetish, and if prompt measures are not taken it will in a short time be completely spoiled. This turning flat and sweetish is due to the mucilaginous substances which, under the action of dilute acids and a favorable temperature, become transformed into substances resembling dextrine and other saccharine matters, which give place, or rather favor, when the alcoholic fermentation has not been of a thorough character, the development of secondary fermentations.

SWEET (*Dolce*, It.; *Doux*, Fr.).—A sweet wine is one in which the sweetness is pleasant, because not excessive, and in harmony with the other principal ingredients, and more particularly with the alcoholic contents.

> " *Il vino dolce e glorioso*
> *Rende l'uomo pingue e carnoso*
> *E allargo lo stomaco.*"

OVER SWEET (*Dolciastro*, It.; *Douceâtre*, Fr.).—This is said of wines which are too sweet, or in which the sweetness does not seem to be well combined; that is, the sugar seems to have been lately dissolved in the wine.

HONEY SWEET, SICKLY SWEET (*Dolce smaccato, Melacchino*, It.; *Doux fade, Mielleux*, Fr.).—Of white wines when they are very sweet and of a nauseating sweetness, resembling must more than wine.

*Melacchino* is perhaps a corruption of *melichino*, meaning cider—*vinum ex malis, pomatium* of the Latins.

NEW OR YOUNG WINE (*Vino giovane, nuovo*, It.; *Vin jeune*, Fr.).—A wine which has been made but a short time, and which has not undergone those changes and transformations in its composition through which it acquires new qualities, due to the new substances which are formed, and which render it more agreeable to the palate, and in the case of fine wines impart bouquet and even sève.

Another cause of variation in the character of wines is the deposition in whole or in part of various substances on the walls of the cask, or in the form of lees at the bottom, that are thus eliminated from the composition of the wine.

These young wines, compared with their condition at maturity, are more heavy bodied, more deeply colored (green or acid), more astringent, and sometimes rough and harsh.

These wines are, finally, more nutritious than after they become mature; it must not be forgotten, however, that a wine which is too young is somewhat indigestible.

GREEN (*Verde, Verdetto, Bruschetto*, It.; *Vert, Aigrelet*, Fr.).—Green wine is not synonymous with young wine, as might be supposed at first; greenness is a quality which a new wine may and generally does have.

A wine is said to be green when it has an acidity and roughness which, though pronounced, is of such a character that it will disappear with time.

Thus, incompletely ripened grapes give a green wine, owing to a small quantity of volatile acid and acid salts which they contain, and more especially bi-tartrate of potash.

Greenness is characteristic of certain new wines, and also of many mature wines produced in northern countries.

TART (*Acidulo, Acidetto*, It.; *Acidule, Aigrelet*, Fr.).—Said of a wine possessing an agreeable and sufficient acidity, due to the presence of free tartaric acid and sometimes of carbonic acid, especially when this latter is in such amount as to become free easily, and so affect sensibly the tip of the tongue.

HARSHLY ACID (*Acerbo*, It.; *Acerbe*, Fr.).—Expresses a sharp, harsh acidity, like that in sour or unripe fruit, which puts the teeth on edge and draws up the lips and mouth. This acidity comes from immature seeds or green stems, which communicate their acids, such as malic, racemic, etc., to the wine; in other words, the acid is the same chemically as that found in unripe fruit.*

Wine produced from grapes which for some cause or other have not reached their maturity, are always more or less harshly acid.

With time this repellant acidity disappears, for the reason, according to Dessaignes, that the malic acid, after eight or ten months, decomposes into succinic and butyric acids.†

MATURE WINE ( *Vino maturo*, It.; *Vin mûr*, Fr.).—A mature wine is one which has quite developed all its characteristic qualities, and which is therefore ready to be drunk, or to be placed in bottles, where, in aging, it will go on improving.

DECREPIT WINE ( *Vino decrepito, passato*, It.; *Vin passé, affiabli*, Fr.).—The caducity of a wine is the stage, according to Dr. Guyot, where it has passed its prime maturity, and when it has already commenced to deteriorate; when, in other words, it has lost some or all of the qualities due to its volatile principles and other constituents.

A decrepit wine has lost its fragrance, has become flat; it has not contracted any disagreeable or repelling flavor, for the taste of *age* that these wines have cannot be called disagreeable in the same sense as a wine which is attacked by the disease called *bitterness*, but it has a slight bitterness which recalls that of some resinous substances.

These wines, when they find themselves in favorable conditions, as when exposed to the air, decompose readily.

"A wine which has been exposed to the cold of winter and the heat of summer acquires in the month of September the taste which Italians call 'settembrino,' which is exhausted and 'passé.'"—M. Salvini.

DRY‡ ( *Vino asciutto*, It.; *Vin sec*, Fr.).—This is said of a wine which leaves in the mouth a sense of dryness. It is a characteristic of highly alcoholic and somewhat astringent wines. "*Pomino* leaves the mouth dry," say the Tuscans. A dry wine is not only without even the slightest taste of glucose, but it does not contain, or only in the most minute degree, the quality of smoothness due to a certain quantity of glycerine, and, in the case of high-class wines, of other substances.

ASTRINGENT ( *Aspretto*, It.; *Un peu âpre*, Fr.).—When the tannin is somewhat noticeable.

---

* This acidity must not be confounded with that due to the acetification of the wine. This excessive acidity may be amended by an indirect method, which is that suggested by Gall, and which aims to correct the must before fermentation. Or some may have recourse to "marmorizzazione;" that is, the addition to the wine of powdered calcium carbonate (marble), which is, however, a method which cannot be very highly recommended, and when necessary, Liebig's method is much to be preferred. This method is to add to the wine a concentrated solution of neutral tartrate of potash in such proportion as to bring down the acidity to the desired degree.

As a preliminary test, to ascertain with an approximation near enough for practical purposes, several quart bottles are filled with the wine to be treated, and to each bottle is added a certain quantity of the solution of neutral tartrate of potash, each bottle being given a slightly greater dose than the one before. The bottles are then corked and left to themselves for a few days. They are then tasted, and the one giving the desired result is used as the basis of calculation for treating the whole quantity.

† The organic acids contained in the must are the following: Tartaric, racemic, malic, citric, tannic, palmitic, stearic, etc.

The acids, on the other hand, which are produced by fermentation, the oxidation of the alcohol, or the breaking up of the sugar, are: Carbonic, acetic, propionic, butyric, valerianic, capronic, œnanthilic, pelargonic, succinic, lactic, etc.

‡ This is a restricted use of the term *dry*, somewhat different from its more general meaning, which is simply *not sweet*, that is, containing no glucose.—*Trans.*

Rough (*Austero, Pavido, Allappante*, It.; *Austère, Âpre, Picotant*, Fr.).—These terms are used of wines which, on account of their excess of tannin, or rather œnotannin, are in the highest degree rough and astringent. Their flavor, which is somewhat nauseous, recalls immediately that of ink, or of ferruginous substances.

In drinking a rough, overastringent wine, a feeling of dryness is produced on the tongue and along the œsophagus. The daily use of wines of this character, by persons of delicate constitution, may occasion organic disorders.

This roughness tends to diminish with time, and may completely disappear; the cause being that the tannin, under the influence of oxygen, gives place to a slow formation of carbonic and gallic acids.

Œnotannin* possesses tonic properties, and insures the conservation of the wine by causing coagulation, and consequently the elimination of many substances which the wine contains, substances whose presence is dangerous from their instability, and because they favor the development of those organisms to which are due secondary fermentations.

High-class and fine wines when young, and even sometimes when old, are more or less markedly rough; this roughness they lose with time.

Harsh (*Duro*, It.; *Dur*, Fr.).—Harsh wines are generally young wines rich in tartar and tannin, and which, consequently, leave a repellant impression on the papillæ of the tongue and palate.

Harsh wines are lacking in delicacy and value.

Harshness, of itself, is a defect; ordinarily it is due to the soil, and in that case the wine is also heavy bodied. This defect may also be owing to unskillful preparation or handling.

Harsh wines keep easily, and can be kept for a longer or shorter time, according to their quality.

---

*Œnotannin has the property of forming with gelatine and with albumen voluminous insoluble compounds, which precipitate with great readiness. By means of clarification, therefore, the contents of œnotannin can be notably diminished, thus curing, or at least considerably lessening, the defect of roughness.

I have called roughness a defect, but that should be understood relatively, not absolutely, for it should not be forgotten that the general trade demands a certain roughness, and wines in which it is lacking are often given this character artificially by the addition of alum, which is undeniably an adulteration, or by the addition of tannin.

Alum is used by unprincipled dealers, and has the quality of reviving the color, precipitating the albuminoids, and imparting a roughness, almost styptic, analogous to that presented by the common Bordeaux wines.

The wine maker has the choice of two kinds of tannin which are found in commerce, and which differ in their mode of extraction or preparation. Thus, the tannin may be extracted from galls by means of ether, giving a tannin pure, but retaining a taste of ether, which renders it objectionable in the treatment of wine. The other kind, which is extracted by alcohol, is inodorous, and therefore preferable for the wine maker.

Pure tannin dissolves completely in alcohol, and in water mixed with 10 per cent of alcohol, and the solution should be limpid. When the wine maker needs tannin he can make use of the grape seeds, which contain a considerable quantity; the seeds may be used either fresh or dry, the latter being more convenient, as they can be preserved from year to year.

It is to be remarked that clarification attempted with isinglass, gelatine, or white of egg, does not always succeed; the failure is due to the lack or insufficiency of tannin in the wine, or to its superabundance.

This explains the common usage of adding tannin to white wines before attempting to clarify them; or in the case of highly tannic red wines why, after adding the clarification, it is often necessary, in order to produce perfect limpidity, to have recourse to sulphuring and racking. This is what the cellarman means when he says that the wine has not taken the finings.

Wines which have fermented slowly, and which contain substances resembling humic compounds, can sometimes be fined even when lacking in tannin.

It is also worthy of remark that tannin has a great influence on the color of wine; it tends to increase it, and, according to M. Nessler, if the wine remains for some time in contact with the lees, it prevents, to a great extent, the diminution of the color.

The life of ordinary or common wines, which are harsh, is limited to a few, two or three, years. These wines in losing their harshness gain little or nothing in value, in fact, as they lose the defect of harshness, they acquire another, that due to tartaric fermentation.

Harsh wines which have good quality and body keep for a long time, and after some years lose their harshness; they thus become more homogeneous, harmonious, and pleasing, or as the experts express it, they become rounded.

If these wines are drunk before they have lost a portion of their harshness, they are not very hygienic.

BITTERISH (*Amarognolo*, It.; *Un peu amer*, Fr.).—This is not a defect; it is even up to a certain point a good quality; that is, when the bitterness is very slight, delicate, aromatic, in short, pleasing; as a rule, a slight touch of bitterness is characteristic of densely colored wines.

Very often this quality is due to the presence of carbonic acid in solution; for example, in young wines or those which have been treated by the Italian method called "il governo."*

Sometimes, in the common language, all wines are called bitter, but with impropriety, which are not sweet; from which the Tuscan proverb, *Vino amaro tienlo caro*, which means, the wine which is not sweet is always of best quality.

BITTER (*Amaro*, It.; *Amère*, Fr.).—Bitterness is a defect, and may be due, as in general it is, to a real malady caused by a micro-organism.

"*L'amertume est pour nous la maladie organique des vins de Pinot.*"— Vergnette Lamotte.

Wines of this kind have a harsh, repelling, nauseating bitterness, due to secondary fermentations, or in the case of young wines, to principles which they have extracted from the skins or stalks during fermentation.

According to M. Nessler the tendency of a wine to this disease is augmented by remaining long in contact with the pomace.

The bitter taste affects principally the posterior portions of the tongue and palate, and the sensation persists for some time.

This fault, which most œnologists consider confined to red wine, is found also, we are told by M. Ottavi, in white wines. He claims to have encountered it in the white wines of Piedmont.

Nessler observes that white wines are less subject to this defect or malady than red, thus admitting, by implication, that they do sometimes become bitter.

The bitter secondary fermentation may develop in any wine, but is more frequent in fine and delicate wines. In common wines the disease usually occurring is the tartaric fermentation.

In general, highly colored wines, rich in extractive matters, are most liable to the attacks of the disease of bitterness.

The high-class wines of Bourgogne, made from the Pinot, not excluding even those made in the most favorable years, are subject to attack by this disease.

In the finest wines Vergnette Lamotte distinguishes two kinds of bitterness: That which attacks the wine during the first two or three years of its life, and which is the most dangerous; and that which shows itself

---

* "Il governo" is a method of wine treatment in common use in Tuscany, which consists essentially in maintaining a slow, protracted fermentation in a poor or neutral wine by the addition of half-dried grapes of high quality, or containing an abundance of those substances lacking in the wine treated, as color, body, tannin, etc.—*Trans.*

in old and decrepit wines. This second bitterness, due perhaps more to chemical reactions than to the action of ferments, is only relatively an ill, as the wine can be consumed before it reaches complete decrepitude.

Pasteur holds that even this second bitterness, which Vergnette Lamotte lays to the account of decrepitude, is caused by the same organism which determines the first kind.

This organism may remain inert for a longer or shorter period, till in the course of aging the wine presents the necessary favorable conditions for its development.

In conclusion, I will say that the bitter taste is a somewhat serious defect; a defect which may be more or less marked, as it may be transitory or permanent.*

---

* The bitter taste in wine may be the consequence of imperfect maturity of the grapes, owing either to an unpropitious season, or to the damage caused by insect or cryptogamic pests; or it may be the consequence of a secondary fermentation, caused by a micro-organism, *i. e.*, the "bitter ferment," which determines the formation of those substances which impart this taste to the wine. In the latter case we have a true disease.

When the bitterness is due to the principles which have passed from the grapes and stems into the wine, then with time and successive finings and rackings it will disappear. This is explained by the supposition that the nitrogenous substances become impregnated with the bitter principles, and thus, when the former are precipitated, they carry along with them the latter, the wine in this way losing this defect.

The bitter taste, if very pronounced, may not disappear after the first rackings, in which case the wine should be fined with gelatine or white of egg.

If the wine be weak, the coagulation of the albumen may be facilitated by the addition of alcohol.

According to the quality of the wine, it may be given a light clarification with the whites of three or four eggs per hectolitre, or a more energetic treatment with 25 grammes of gelatine.

Such a treatment not being found sufficient, recourse must be had to the use of olive oil of good quality; of this the dose to be used is one half litre per hectolitre. The oil is poured into the wine, the whole thoroughly stirred, and then allowed to rest; the oil separates from the wine, and carries with it the substances which have caused the bitterness.

Directly after racking a wine with access of air, it will sometimes become slightly bitter; this seems to be caused by the action of the oxygen of the air upon substances contained in the wine; later the bitterness disappears, owing very probably to the rapid oxidation which causes these substances to precipitate. In this way M. Mona explains how bitter wines in bottles can, with time, lose this defect.

Formerly various opinions were held regarding this malady, because, in all probability, people failed to distinguish between bitterness proper and the malady due to tartaric fermentation, or "la pousse."

Thus De Blassis attributed it to changes of the salts, especially of bi-tartrate of potash; Machard to an invisible action of the fermentative principle, decomposing the last remnants of sugar and salts in the wine; Lebœuf to an abnormal fermentation, which produced, sometimes, citric ether, which has a bitter taste; Vergnette Lamotte to a secondary fementation, caused by a parasitic vegetation, which decomposed the wine in consuming the coloring matter; Neubauer found that the quantity of tannin and of coloring matter diminished with the progress of the malady. Finally Pasteur, after the study of many bitter wines, has demonstrated that this malady is caused by the action of a micro-organism, which multiplies with extraordinary rapidity in the superior wines of the "Côte d'Or," but very slowly in the common wines of Bourgogne, the Jura, and the Bordelais. He adds that this malady presents many diversities in its development, according to the origin and the nature of the wine, but that all wines are subject to it.

Ducleaux, in 1873, determined the volatile acids of bitter wines, the following being the result of his analyses:

| | Volatile Acid. | Total Acidity. | Acetic Acid. | Butyric Acid. |
|---|---|---|---|---|
| Sound wine | 1.01 gr. | 4.40 gr. | 0.97 gr. | 0.04 gr. |
| Bitter wine (1866) | 1.50 gr. | 5.15 gr. | | |
| Bitter wine (1873) | 1.95 gr. | 6.67 gr. | 1.83 gr. | 0.19 gr. |

The increase of total acidity in the sick wine being greater than could be accounted for by the formation of acetic acid at the expense of the alcohol, it must be attributed to the fermentation of the glycerine, which, in fact, had diminished.

EARTHY TASTE (*Terroso*, It.; *Terreux, Goût de terroir, Goût de pièrre à fusil*, Fr.).—By the term earthy a single definite taste must not be understood, but divers flavors which are all in general disgusting or bad.

In tasting, these flavors are perceived by the posterior part of the mouth, and may have their origin in the soil, in the use of inappropriate fertilizers, in the plants supporting the vines, or in the weeds infesting the vineyard, etc.

"The *earthy taste* is a vague term," writes Ottavi, and with justice, for it is a taste which is not always very definite, resembling sometimes earth, manure, flint, slate, nuts, willow, grass, etc. It is well known that Aristolochia, Mercurialis, etc., if allowed to grow in the vineyard, communicate their flavor to the grapes, and therefore to the wine. Pliny was not mistaken when he wrote: "In general, the vine takes up with an astonishing facility the flavors of neighboring plants. The grapes grown in the marshy soils of Padua have a taste of willow."

Generally the earthy taste is not found in high-class or fine wines. I say generally, because there are exceptions; for example, Chablis has a slight flavor of flint, and yet it is a wine of a certain renown.

Richelieu, speaking to Louis XV of a certain wine of Graves, said: "*Il sent la pièrre à fusil comme une vieille carabine.*"

The *flinty taste*, writes Petit Lafitte, has something vinous and energetic, which exactly recalls the sensation experienced by the olfactory organs when a flint recently struck by the steel is held under the nose.*

---

The diminution of the glycerine was also pointed out by Pasteur, who, besides, stated that the tartaric acid did not diminish.

As the researches of Fritz have shown, many microbes are able to cause fermentation of the glycerine ; thus, under the action of the *Bacillus butylicus* it is transformed into butylic alcohol and butyric acid.

Recently, B. Hass experimented with a view of ascertaining whether the bitter taste was due to citric ether, as Müller and other French chemists had supposed, or to some resinous substance produced by changes of the aldehyde in presence of ammoniacal compositions having their origin in the albuminoid matters of the wine.

By exhausting a wine which was afflicted with the *bitter* disease, and which he had previously rendered alkaline with ether, he obtained a resin slightly soluble in water, very soluble in alcohol and in acetic ether, insoluble in carbon bi-sulphide, turning brown in contact with the alcohols, becoming greenish with ferric chloride, and having the extremely bitter taste of the diseased wine.

Hass has found by his experiments that the best way of curing a wine afflicted with this malady, is by the use of oxidizing agents. Oxigenated water in small quantities is inefficacious; in larger quantities it destroys the bitter taste, but produces another not less disgusting. The best results have been obtained by aeration.

The wine is fortified by the addition of alcohol till it contains 13 per cent by volume, if of feeble character and liable to acetify. A current of air is then passed through the wine for two hours, and the bitterness disappears completely.

Filtration through pomace or cellulose has an excellent effect, the bitter substance seeming to be removed by physical attraction.

This disease may be said to have several stages. At first the wine is still clear, but less fragrant, duller in color, and with a slight bitter taste. Later it acquires an odor *sui generis;* the bitter taste increases, becoming piquant on account of the small quantity of carbonic acid produced by the secondary fermentation which takes place. Finally it loses its natural color, becoming brownish, with a tendency to blue; there has then taken place a serious change in one of the principal components of the wine—the extractive matter—and the wine has become an undrinkable liquid.

* According to Doussieux, the earthy taste is due probably to the solution and evaporation of a part of the mineral and metallic substances which are found in the soil of certain vineyards.

Petit Lafitte seems inclined to attribute the flinty taste to iron and alumina.

Ladrey, on the other hand, accounts for it by the presence of much silica in the soil, and many analyses show silica not only in the leaves and seeds of the vine, but also in the wine.

Joulie states that the flinty taste is due to the fact that pyroniac silica contains a bituminous substance of organic origin, the peculiar taste of which is communicated to the wine.

It should also be remembered that the experiments of Thenard prove that silicate of lime is much more soluble in water than was formerly believed.

According to the experiments made by Aubergier, the principle to which wines owe their earthy taste is found neither in the seeds nor in the stems, but in the skins of the grapes. From 15 kilogrammes of pomace he extracted 30 grammes of a volatile oil so acrid and penetrating that a single drop was sufficient to infect 10 litres of the best brandy.†

This fact supports the opinion of those who see in the prolonged contact of the wine with the pomace the cause of the earthy taste.

Certainly, by improving the soil, by the use of proper fertilizers, by a good defecation of the must, by a prompt removal of the wine from the pomace, by clarification and rackings, the taste under discussion is much diminished, and sometimes completely eliminated.

TASTE OF SOIL (*Sa di terra*, It.; *Goût de terre*, Fr.).—When the wine has that taste of soil or of clay, due to the presence of soil in the must during fermentation. The soil in the must may come from the skins of the grapes, which may easily become covered with it when the bunches lie too close to the ground, or may have become mixed with the grapes accidentally or by carelessness.

This taste may come, also, from the clay which the peasants sometimes use as cement to close the leaks in tubs, vats, or other utensils.

TASTE OF BRINE, SALT (*Sa di salmastro, di salso*, It.; *Goût de saumâtre, de salé*, Fr.).—The wine has sometimes the taste of common or culinary salt.

This defect is found in wines grown in soil rich in salt, or in localities near the sea.

COOKED TASTE (*Sa di cotto*, It.; *Goût de cuit*, Fr.).—If the wine has a taste more or less pronounced of must or caramel, due generally to the action of fire upon the must when the latter has been concentrated carelessly, or by direct heat.

This taste is caused, also, by an over-maturity of the grapes, as happens in very hot weather, and especially when the grapes are thick-skinned; it may be caused, also, by frozen grapes, or by the freezing of the wine; in the latter case especially when the pieces of ice formed in the wine are not carefully removed.

RESINOUS TASTE (*Sa di resina*, It.; *Goût de resine*, Fr.).—This taste is found in wines which have been kept in receptacles made of resinous wood.

BREAD TASTE (*Sa di pane*, It.; *Goût de pain*, Fr.).—Some sweet liquor wines have an agreeable taste which reminds one of the odor of fresh bread.

TASTE OF DRUGS, MEDICINAL TASTE (*Sa di droghe*, It.; *Goût de drogues*, Fr.).—A taste due to the addition of some infusion or drug to the wine.

---

Regarding the quantity of silica contained in wine, we have the analyses of Boussin-gault, who, in analyzing his wine grown at Smalzberg (Bas Rhin), found 6.096 gr. of silica per 1.870 gr. of ash in a gallon of wine, 5 per cent of the mineral ingredients.

Grasso, in the ash of four different musts, found the following quantities of silica:

| | |
|---|---|
| Petit Bourgogne (not mature) | 1.991 per cent. |
| Petit Bourgogne (mature) | 2.099 per cent. |
| Petit Bourgogne (mature, but from a different soil) | 1.191 per cent. |
| Grün Sylvaner (white, mature) | 2.181 per cent. |

In the skins the proportion was greater; in those of the first it was 3.464, and 2.571 in those of the fourth.

† That a drop of this oil is capable of infecting so large a quantity of brandy is not wonderful, when we reflect on the sensibility of our organism, especially of our sense of smell, which is so susceptible as to surpass the extremely delicate spectroscope. Thus, for example, Valentin has shown that one five hundred thousandth of a milligramme of sulph-hydric acid, or one two millionth of a milligramme of essence of roses, is sufficient to make an impression on our olfactory organs.

BURNT TASTE (*Sa d'abbruciato*, It.; *Goût de brûlé*, Fr.).—When the wine has a flavor of acrid fruit, together with a spurious cooked taste.

The taste of which we speak is a consequence of the partial withering of the grapes before their maturity, on account of extreme heat or of great changes of temperature between night and day.

MOUSEY TASTE (*Sa di topo*, It.; *Goût de souris*, Fr.).—A wine will sometimes have a disgusting flavor and odor that recalls forcibly the odor of the excrements of mice. The cause of this defect is not well known. According to some authorities, it is due to lack of cleanliness in the receptacles in which the wine is kept. Others believe it to be caused by the action of the oxygen of the air on the extractive matter of the wine, for there seems sometimes to be a distant analogy between the mousey taste and the fresh bread taste so much appreciated in some liquors. It is very probable that both of these causes concur to produce this taste, for it is found sometimes even in wines which have been kept in glass.

The mousey taste may be more or less intense, and wines affected produce a dry feeling in the mouth when they are tasted. If a wine has this taste in a very slight degree it is not noticed immediately; it often happens that after passing judgment on a wine, one's opinion has to be modified by a mousey taste which is not perceived at first. If the defect is pronounced, it is perceived immediately by the nose; the odor and taste too, in this case, are so disgusting as to be sickening.

HEATED TASTE (*Sa di riscaldato*, It.; *Goût de réchauffé*, Fr.).—This unpleasant flavor is hard to define, as, in fact, it is a mixture of various flavors—of acetic acid, of stems, of organic matter slightly decaying under the influence of heat and moisture, etc.

This taste is easily produced by allowing the cap to become overheated during fermentation, or by heating grapes before crushing them.

With time this taste tends to disappear, but when somewhat pronounced it diminishes, leaving the wine with a somewhat acrid taste.

SULPHUR SMELL, or better, SMELL OF SULPH-HYDRIC ACID.—An odor resembling rotten eggs which a wine may have, and which is due to the presence of sulph-hydric acid or suphuretted hydrogen.*

TASTE OF STALE EGGS (*Sa di uova stantìe*, It.; *Goût d'oeuf gaté*, Fr.).— This taste, which is easier to avoid than to cure, comes from the use of eggs not perfectly fresh for fining.

ODOR OF SULPHUROUS ACID, OR OF SULPHUR VAPOR.—A wine often has the odor characteristic of this substance when it has been recently racked into an excessively sulphured cask.

As every one knows, things that are useful when used in moderation become dangerous when used in excess. This is the case with sulphurous acid.

---

* It is generally held that the cause of the formation of sulph-hydric acid in the wine is the presence of sulphur in the fermenting mass, as happens when the vines have been sulphured in such a way as to allow sulphur to adhere to the grapes. This is indeed the principal cause, but not the only one. Nessler cites six of these causes, which are: The sulphuring of the vines; the sulphuring of casks; the use of sulphur tape; the use of certain fertilizers; the cultivation of the vines in certain soils; the presence of iron in the vats or casks.

To these causes, most probably, should be added another, that of the reduction of sulphates by micro-organisms, a reduction first noticed by Planchud, who attributed it to vital action. This action has been found by Etard and Olivier to be due to algæ of the group of oscillators, called *Beggiatoa* (*B. roseo-persicina, B. mirabilis, B. alba*). Other algæ of the genus *Ulothrix* have the same property.

Is it not possible that micro-organisms might be found in wine resembling and acting in the same way as these algæ found in sulphurous waters?

The fine experiments of Dubœuf and J. Bruhl on the action of sulphurous anhydride, or acid, on micro-organisms, have an important bearing here.

They have deduced from their experiments the following conclusions:

1. Sulphurous acid gas has an evident microbicidal action on the germs contained in the air.

2. This action is especially perceptible when the air is saturated with water vapor.

3. Sulphurous acid acts particularly on the germs of bacteria.

4. Pure sulphurous acid will destroy germs, even in the dry state, if the action is sufficiently prolonged.

Sulphurous acid, when used in excessive quantities, takes away from the quality and color of the wine, and gives it a bitterish, astringent, and displeasing taste. In time the sulphurous acid changes to sulphuric, and then into sulphate of potassium. This is why in many wines is found a certain quantity of this sulphate, which is dangerous to health, and, when sufficient of it is present, would lead to the belief that the wine had been plastered.

At the end of the last century it was shown that a wine sulphured to excess acquired a very disagreeable odor, and was hurtful to the health, causing headache, vertigo, oppression of the stomach, nausea, etc.

In practice it is good to remember that the more alcoholic a wine the more sulphurous acid it will dissolve or absorb.

Nessler, making a comparison of water and wine at 9 per cent of alcohol, filled a barrel quickly with each, after having burned as much sulphur as the air in the barrel would consume, and found that the water absorbed .01035 per cent of sulphurous acid, and the wine .01346 per cent.

The quantity of sulphurous acid which a wine will absorb in process of keeping cannot be exactly stated, as it depends on the number of sulphurings, the amount of sulphur burned, or, when the sulphur is burned directly in the cask, on the amount of oxygen there.

According to Weigert the quantity of oxygen in a cask of one hectolitre is 21 litres or 30 grammes. By burning an equal quantity of sulphur 60 grammes of sulphurous acid are formed. When the cask is filled all this is not dissolved, because part is oxidized immediately, and part escapes into the air as the wine enters the cask; thus, the total amount absorbed by the wine is reduced to about 10 or 11 grammes.

VARIOUS ODORS ( *Violet, Rose, Mignonette, Pink, Bitter Almonds, etc.*).— These are all odors given artificially to the wine to render it more fragrant, or to attempt to pass it off as a wine of higher quality than it really is.

Many high-class and fine wines, in aging, develop characteristic bouquets; but besides bouquet these wines have sève, which artificially perfumed wines lack altogether or have little of in proportion to their fragrance.

Besides the odors which we call good, which have been added artificially, we have also bad odors which are absorbed from the air by the grapes or the wine, such as the odor of tobacco, of grass, etc.

WOOD TASTE ( *Sapor di legno, Asciutto, Sa di secco*, It.; *Saveur de bois, Seche, Goût de sec*, Fr.).—A taste not easily defined, as it lies somewhere between that of wood and of mold. It is communicated to the wine by ill-kept casks which have become " secco, asciutto," a defect seeming to

4

be due to the development of mold in the inside of the cask. Sometimes wine will acquire this taste when left long with ullage or in imperfectly closed casks.

To remove this taste recourse is had to olive oil, lemons, or refermentation with a small quantity of fresh grapes.

" *Se egli sappia di secco, il vino, vi abbia odor cattivo, caccinvisi dentro fiaccole acuse, e vi si spengano.*"—Soderini.

Taste of the Stems.—This is a rude, unpleasant taste, vulgarly known as a taste of "legno verde" (green wood). It is found in wines which have been allowed a too prolonged contact with the stems, or which have been made by a maceration of the whole bunch, or which have been made from bunches not perfectly sound. The taste of stems is generally accompanied by some bitterness.

Clarifications and rackings with contact of the air will often destroy or notably diminish the stem taste.

When it is desired to prolong the contact of the wine with the pomace, stemming is to be recommended.

Smoky Taste.—This taste resembles the smell of burning wet or green wood. It is, writes Mona, somewhat acrid and bitter, recalling smoke and soot. According to Mona, it is found more rarely in Italian wines than in German.

This defect may be occasioned by the smoke given off by ill-constructed stoves used to heat the fermenting-room or cellar; or it may be due to unfavorable climatic conditions during the vintage.

It has been stated that musts corrected by the addition of cane sugar will sometimes give wines with this taste.

With the smoky taste a wine loses its brightness, becomes cloudy, and if not cured by sulphuring, changes into a liquid not to be tolerated by even the most uncritical palate.

Oak Taste.—A taste which a wine will contract after two or three rackings into new casks which have not been properly prepared, especially if they are made of a bad quality of wood. The wine in this case acquires a peculiar, bitterish taste, according to Ottavi, almost aromatic, much tannin, and often the real flavor of the wine is quite destroyed.

Taste of Mercaptan.—The repugnant taste and odor of onions or garlic, which remains even after the wine has been racked into well-sulphured casks.

The same causes which tend to produce hydrogen-sulphide in the wine, not excepting plastering when it is done heavily, tend also to form mercaptan. So far no means have been discovered of removing this taste from wine.

Polacci was the first to observe the formation of these products, which have a fetid and persistent odor, and are due to the action of sulph-hydric acid and sulphur on the components of the must and wine; he believes them to be simply ethylic mercaptan. Konig thinks that this reaction is not very probable, as it has never been known to take place in a dilute acid solution. He believes, on the contrary, that the aldehyde contained in most wines combines easily and directly in a dilute acid solution with sulph-hydric acid to form thio-aldehyde and trithio-aldehyde. Now these compounds are endowed with a strong, persistent, and disagreeable odor, resembling closely that acquired by wines containing sulph-hydric acid; it may be, therefore, that the mercaptanic substance spoken of by Polacci is nothing but thio-aldehyde or trithio-aldehyde.

TASTE OF LEES.—Wine, by a prolonged contact with the lees, loses its clean taste and acquires a more or less pronounced bitterness, which has a distant resemblance to a taste of decay, and is characteristic of lees even when sound.*

TASTE OF DECAY (*Sapore di fradicio*, It.; *Saveur de pourri*, Fr.).—A taste which the wine contracts from unsound cooperage or too prolonged contact with the lees; it is a repelling taste of rottenness, which, however, must not be confounded with that caused by putrid fermentation of the wine.

This taste may also originate in imperfectly ripened grapes, which, through the prolonged action of dampness, have commenced to decay.

If the grapes are ripe before they commence to decay, the wine will still have something of this taste, but it will be less disgusting and will tend to disappear with time; the wine will, however, always be insipid, and lack frankness of taste.

MOLDY TASTE.—The characteristic taste of mold. Wines easily contract this taste, either from moldy casks or from moldy grapes having been used. It is generally possible to take away this taste by the use of olive oil.

*Sapore di tempesta*, It.; *Saveur de grêle*, Fr.—A harsh, bitterish, somewhat moldy taste, perceived in wine made from grapes that have been injured by hail at the commencement of their ripening.

RANCID (*Rancido*, It.; *Rance*, Fr.).—"When the wine is swallowed, or whilst it is being drunk, a displeasing taste is noticed in the throat and slightly on the palate, almost analogous to that of rancid substances, from which comes the name given to this disease of wine, till now unstudied by any author. The *rance* can also be smelt, if it is pronounced, but a good nose is needed to discover it, and a delicate palate to taste it, at its incipiency."—O. Ottavi.

FRUITY TASTE† (*Sapore di frutto*, It.; *Saveur de fruit*, Fr.).—Many young wines, when well made, have a very pronounced taste of fruit.

Common wines, with age, lose this taste, but fine, and above all, the finest, wines retain it, much to their advantage; they retain it, however, only when aged slowly, and without the use of artificial aids.

TARTARIC FERMENTATION.—This term is used to cover two different maladies of wine caused by two micro-organisms, which differ somewhat from each other, and the products of the fermentations caused by them differ considerably. These maladies, however, have a certain affinity, since both the micro-organisms, to whose action they are due, live at the expense of the tartaric acid in the cream of tartar.

The French distinguish these two maladies, calling the first "*la maladie de la pousse—vin poussé;*" in Italian, "*malattia del subbollimento;*" and the second, "*maladie de la tourne—vins tournés;*" in Italian, "*cercone.*"

---

* It may perhaps be useful to note that the lees may become the seat of a bacteroid fermentation independently of any anterior disease in the wine. Thus, according to the experiments of Ravizza, the wine and lees may become the prey of bacteria without the aid of molds or other micro-organisms that destroy the acids.

The temperature most favorable to the development of bacteria in the lees seems to be from 77° F. to 86° F. Below 77° F. the phenomena accompanying the life of these bacteria decrease, and towards 50° F. cease altogether. The practice, then, in racking, of separating the last layers of wine, that is, the part lying in contact with the lees, from the rest is a good one, and this wine may be considered of inferior quality, either because it lacks a clean, fresh taste, or because it is sometimes cloudy.

† Fruity is very often used in English with the inappropriate meaning of somewhat sweet.—*Trans.*

"*Maladie de la pousse.*"—This disease is recognized by the wine spurting out when the vessel in which it has been confined is opened; the wine exercises a strong pressure on the staves of the cask on account of the carbonic acid which is formed; it is from this that comes the term "pousse."

In the glass the wine shows a persistent ring of small gaseous bubbles of a whitish color. If the wine is left exposed to the air it becomes turbid; its color becomes dull with a tendency to yellowish.

The wine has lost its primary flavor, and as the disease progresses, becomes more and more insipid; if it is shaken there is an appearance of silky waves at the surface, caused by the lees which has risen up.

Balard was the first to show the presence in "vins poussés" of a ferment which, according to him, resembles the lactic ferment. He has further shown that in these wines the quantity of volatile acids is increased, the one found in largest quantity being acetic acid.

Bechamp and Sténard have shown that propionic acid is formed in these wines from the tartar and the glycerine. Nicklés, on the contrary, is of the opinion that metacetic acid is produced.

Duclaux, who has given much attention to this malady, seems to have proved: (1) That the amount of free acids augments with the progress of the malady; (2) that this increase is made at the expense of the fixed acids of the wine, particularly of the tartaric acid; (3) that the acids formed are propionic and acetic. After having shown this he concludes by saying that all fermentation of the tartar that takes place with the evolution of pure carbonic acid and production of propionic and acetic acids should be called "*maladie de la pousse.*"

*Cercone, vin girato, mercuriella,* It.; *Tourné, vin tourné, vin qui a donné le tour,* Fr.).—At this word in an Italian dictionary is written: *Cercone* —a distiller's term—is said of a spoiled wine, because in becoming thus it works and turns; *vappa, lora* of the Latins. The *lora* of the Latins is certainly not the *cercone,* but *family wine, piquette;* neither is *vappa,* since that, according to the dictionaries, should indicate a flat, vapid wine. *Vappa vinum insipidum et nullino virtutis, postquam omnino odor saporque optimus evaporavit.*

*Vin tourné* has this peculiarity, that when first poured out it appears sound, but after a short time it tends to become turbid and iridescent.

Under the influence of the oxygen of the air the coloring matter becomes purplish, and precipitates, and the wine acquires a yellowish tint, a sour taste, and a forbidding bitterness.

Wines of this kind when distilled give a brandy having a bitter taste, caused probably by ammoniacal compounds. The alcohol made from them has not always, but often, a strong and pungent odor, and cannot, without being well rectified, be put to the ordinary uses of wine alcohol, that is, the manufacture of vermouth, etc. This odor is sometimes so pungent as to bring tears to the eyes, and, by fractional distillations, it is possible to isolate a certain quantity of croton-aldehyde.* This compound is formed, very probably, during the distillation by the condensation of the aldehyde with diminution of water.

Balard has found lactic acid in "vins tournés;" Glenard, on the other

---

* Recently Professor Comboni, in distilling a wine made by blending Marzemino and Black Pinot, which had been attacked by the bitter fermentation, found in the distillate a considerable amount of aldehyde and formic acid. These products are certainly formed during the progress of the secondary fermentation, for they are not found at all in the same wine when sound.

hand, has found potassic acetate. In the secondary fermentation of "vins tournés," there is a formation of acetic acid, and more especially of lactic and tartronic acids.

A wine attacked by this disease may be considered as lost; however, at the start it may be useful to try the addition of tannin and cream of tartar, then pasteurization and fining. The disease, if not arrested, is followed by putrid fermentation.

PUTRID FERMENTATION.—This disease attacks the organic matter in the wine, destroys it, and gives rise to repulsive tastes and odors.

In the incipiency of this fermentation, the repulsive odor and taste are not very marked, and a cure may be attempted by heavy sulphuring, followed by filtration through charcoal, which acts as a disinfectant.

FAT (*Grasso*, It.; *Gras*, Fr.).—I will say now that this defect should not be confounded with that of viscosity or greasiness, though at first view it might be supposed to be the same in a moderated form.

The defect of "fatness" is rarely found in generous wines, but is usually confined to weak ones, and is not due, like "greasiness," to a fermentation, but to the presence of a certain amount of albuminoid substances, of gum, mucilage, imperfect sugars, etc., which impart to the wine a character which, when it is tasted, leaves a more or less marked impression of something glutinous; an impression which persists for some time, leaving, as it were, a pasty feeling in the mouth.

"Fat" wines are indigestible, and hard to keep during the hot season, as they are extremely liable to secondary fermentations. The wines in which this defect is usually found, are those grown on moist plains, which are naturally fertile, or made so by the addition of nitrogenous manures, as, for instance, young vineyards where the effect of manuring at the planting of the vines has not worn off.

This defect may be avoided entirely, or to a great extent, by a thorough and prolonged aeration of the must, or by the addition of alcohol or tannin* to the wine.

Sometimes this defect, when not too pronounced, will partly or wholly disappear after the wine has gone through its slow spring fermentation.

GREASY, VISCOUS (*Filante, Grassume*, It.; *Filante, Graisse*, Fr.).— Terms used of a wine which has lost part of its fluidity and which, when poured into a glass, falls without noise, or like oil; it has that viscid, mucilaginous look which reminds one of white of egg.

This malady is caused by a micro-organism. A greasy wine, as the malady progresses, loses its fragrance and becomes bitterish; its color becomes dull and tends to turn brown; finally, it loses its natural transparency and brightness. At first it is flat, vapid, and distasteful; and finally, rancid and sour by the formation of lactic acid.

---

* The addition of a little tanninized wine is better than the direct addition of tannin. Tanninized wine may be prepared thus: Take a small cask, holding, for example, about 25 gallons; fill it with a strong wine, or one made so by the addition of 1 or 1¼ gallons of alcohol of 94° C.; into the wine put about 35 pounds of grape seeds which have not been fermented. For the first few days the wine should be stirred from time to time, and then left to itself. After about ten days the liquid part is drawn off, and is then a wine heavily charged with tannin, which serves excellently for the purpose above noted; for that purpose a dose of 1 or 2 gallons of the tanninized wine to 100 of the wine to be treated is about the right proportion.

If a tanninized wine is needed for the defecation of the must, it is prepared thus: Take 5 gallons of alcohol and 10 gallons of wine, put in a small cask, and add about 18 or 20 pounds of seeds, and treat as in the former case; 1 or 2 gallons of this is sufficient to thoroughly defecate 100 gallons of must.

If fresh seeds are not to be had, dry ones may be used, providing they are in good condition, that is to say, providing they have been dried in the shade, kept in a dry place, and have not become moldy.

This malady occurs oftener in white than in red wines; in late years, however, it has been found often in red wines on account of the unfavorable conditions for the grapes attaining a complete maturity, such, for instance, as the damages done by insects, cryptogams, and bad weather. It occurs easily, too, in red wines made from grapes grown on very fertile soil rich in albuminoid substances.

Very probably this deterioration is much more complex than is usually supposed.

Peligot was the first to establish the presence of a micro-organism, of a bacterium. Pasteur, later, demonstrated that this bacterium has the property of transforming the sugar that remains in the wine into a mucilaginous or viscid substance.*

Béchamp calls the active ferment of this process *Micrococcus viscosus*, and the gum which is formed *viscosio*.

Tannin and alcohol, in certain proportions, prevent the development and action of this bacterium; the conclusion, therefore, is that wines poor in alcohol and tannin, and containing sugar, are subject, especially if white, to become "filant." This explains also the use of tannin, as proposed by M. François, of Chalons, to arrest or prevent this malady.

François attributes this malady to a peculiar nitrogenous substance, gliandin, a kind of glutin, which seems to have the property of being precipitated by tannin. Nessler affirms, however, that we do not know yet how the tannin acts.

I have already remarked on the complex nature of the malady under discussion. Usually it is held to be owing to a lack or deficiency of tannin. This, however, is not invariably true, since Francisco Selmi has found it in wine made from Lambrusca grapes, and therefore rich in tannin. It seems that in this malady the tartaric acid also suffers changes. Probably on account of these changes Bizzari proposes the use of tartaric acid, 200 to 250 grammes per 100 gallons, as a cure or preventive of the malady.

---

*The bacterium of "La Graisse" put into a solution of sugar containing albuminoid and mineral substances acts upon the sugar and transforms it into a kind of gum, mannite, water, and carbonic acid. Thus, 100 parts of cane sugar will give 50.09 parts of mannite, 43.5 of gum, besides water and carbonic acid.

Monoyer proposed to account for this transformation by two chemical equations, the first of which would give mannite and carbonic acid, the second gum and water, as formed from the glucose.

Schmidt-Mülheim is about of the same opinion, he believing that the viscous fermentation consisted of two processes, the first of which gave mannite and carbonic acid, and the second the viscid substance.

Kramer has studied this ferment. He examined three wines afflicted by it, and besides *Saccharomyces ellipsoideus, Saccharomyces mycoderma*, etc., he found an extremely minute bacillus 2 to 6 $\mu$ long, and .6 to .8 $\mu$ thick. He failed to cultivate this bacillus on potato, agar agar, etc., but by putting a little of the infected wine into a new (three months) white sterilized wine and with 3 per cent of glucose, he found that the bacillus developed well and rendered the wine "filant," but only when the air was completely excluded by covering the wine with a layer of oil. With access of air there was very little development of the bacillus, and instead an increase of the other ferments of the wine. Kramer has called this ferment *Bacillus viscus vini*.

The peculiar kind of gum produced by the viscous fermentation of the sugar renders the wine viscid and glutinous. In its properties it resembles dextrine more than it does gum arabic.

The viscid substance, according to Kramer, appears to be a product of assimilation of the organism, whilst the carbonic acid and mannite, which are formed contemporaneously, are products of the fermentation; a constant proportion between the first and the last substances does not exist.

The bacillus multiplies very well in its own viscid product.

The gum can be isolated and purified by precipitation with alcohol, dissolving the precipitation with water, and re-precipitating with alcohol. Dried at 100° C., it forms a brown, amorphous body, which in water, without being dissolved, swells up greatly and forms a kind of glue. It has no acid reaction.

The best means of preventing or arresting the disease consists of the use of tannin, pasteurizing to destroy the bacteria, racking into sulphured casks, and finally the addition of alcohol to the wine.

Pasteurization is inapplicable in the case of white wines which are destined for the fabrication of champagne, because it not only destroys the bacteria, but also the alcoholic ferments, whose action is necessary to produce the carbonic acid, which renders the wine sparkling.

At the beginning of the development of the disease, forcible agitation of the wine will restore its clearness and fluidity by the disassociation of the bacteria and the dispersion of the mucilaginous matter which envelops the parasite.

Agitation, however, must not be looked upon as a curative measure; the results obtained are only temporary, for the cause of the disease, viz.: the bacterium, is neither destroyed nor removed.

FLAT, WINE FLOWERS (*Vino svanito, Svaporato, Fiorito*, It.; *Vin evanoui, Évaporé, Fleuri*, Fr.).—A wine becomes flat when it remains for some time exposed to the air, as happens in an imperfectly filled or badly bunged cask. In time it becomes covered with "wine flowers," which consists of the *Saccharomyces vini*, or *Mycoderma vini*. In either case the wine gradually acquires an unpleasant, somewhat bitterish taste, and loses its strength and bouquet by evaporation, or else the breaking up of the alcohol into water and carbonic acid. This has been called by some one, on account of the products formed, hydro-carbonic fermentation, and is caused by the *Mycoderma vini*, which attacks not only the alcohol, but very probably the ethers, succinic acid, and glycerine, as these bodies tend to disappear.

Although cases do occur in which generous wines are attacked by the *Mycoderma vini*, still it has a decided preference for young and feeble wines. In old and well-defecated wines it develops with difficulty, perhaps because in these wines the elements necessary for its nourishment (nitrogenous bodies and phosphates) are not found.

The practice of some wine makers with regard to "wine flowers" is not in accord with that of those who follow a rational system of wine making. They consider only the development of the "flowers," which they look upon almost as a preservative of the wine, whilst the others sustain the necessity of energetically combatting and preventing the increase of the "flowers," because it is not only dangerous in itself, but is almost always accompanied by the *Mycoderma aceti*, or *Diplococcus aceti*, which, the moment circumstances become favorable, commence to replace the *Mycoderma vini* and cause the acetification of the wine.

When it is thoroughly understood how the "flowers" act it is easy to explain the facts put forward by those who do not consider it dangerous, and also the reasons of those who believe that it should be prevented by all means, and destroyed on its first appearance.

The presence of the "flowers" causes such an absorption of oxygen and development of heat and carbonic acid, as to prevent the growth of any other organism.

Ducleaux has calculated that 80 grammes of alcohol contained in a litre of wine of 10 per cent, needs for its transformation into water and carbonic acid more than 160 grammes, or 100 litres of oxygen.

The conclusions to be drawn from this are evident; they are, that when the cask is well closed, so as to prevent the free entry of air, the diminution of alcohol, caused by the "flowers," is reduced to a mere trifle, and that the presence of the "flowers" excluded the action of other micro-organisms.

We must not, however, reason from this that the *Mycoderma vini* is really of use, for if exposure to the air should happen, if, instead of remaining white, the "flowers," as Pasteur noticed, should turn red, then, sooner or later, it will cede its place to other organisms, to the vinegar *diplococcus*, which, as I have shown before, is ready immediately to commence action, finding itself in favorable condition for its development, for the "flowers" itself serves for nutriment; and if there should be a considerable rise in temperature, the conditions are the best possible.

The final conclusion then, plainly is, that the "flowers" should be carefully guarded against; this is done by the strictest attention to "filling up," the importance of which was recognized by the poet Alemann, when he wrote:

*Che nulla cosa*
*Può medicar il vin, che resta scemo.*

The "flowers" may be destroyed by the addition of sulphurous anhydride or a few drops of alcohol.

With wine in bottles, the development of the "flowers" is prevented by keeping the bottles lying down; if instead the wine is kept in flasks ("*fiaschi*"), as in Tuscany, or in demijohns, a few drops of the purest olive oil on the surface of the wine will have the same effect.

SOUR, PRICKED, ACETIFIED (*Vino che ha preso il fuoco, Lo spunto, La punta, Il portore, Vino acetoso*, It.; *Vin qui a pris le feu, l'Aigre*, Fr.).— Acetic acid is one of the normal components of wine. It is formed during the alcoholic fermentation, but in such minute quantities as to be imperceptible to the taste. When the proportion of this acid, from one of the many known causes, becomes large enough as to be perceptible, then the wine is said to be "pricked."

A pricked wine retains its natural color and limpidity.

This defect is recognized by the odor and taste of acetic acid; in tasting, its strongest effect is perceived at the base of the tongue.

If a wine thus affected is not taken in hand immediately (and in truth success is not always sure) and treated with heavy sulphurings or pasteurizing, it soon becomes sour and acetic.

Acetification is due to the action of a micro-organism, the bacterium known under the name of *Diplococcus aceti*, still commonly called *Mycoderma aceti*, which increases with a rapidity truly prodigious. Ducleaux tells us that if on a surface of wine a metre square an almost imperceptible amount of these bacteria is allowed to fall, in twenty-four hours the whole surface of the liquid will be covered with a layer of them so closely placed as to be crowded into contact. Thus, there will be three hundred thousand million individuals formed in twenty-four hours.

The rapidity with which the acetic bacterium multiplies explains why a pricked wine, when the temperature is favorable, becomes so quickly completely acetified.

It should be remembered that whilst it is easy to prevent this disease by taking proper precautions in the fermenting-room and cellar, it is difficult, if not impossible, to destroy it when started.

Once a wine has become pricked, instead of trying to effect a cure, it is better to follow the advice of Guyot, who says:

"When wine acquires the odor and taste of acetic acid, it is sent to the vinegar factory, but it is never attempted to use it as wine."

All the means that have been suggested for the treatment of a pricked wine may be considered as palliatives only, and not as radical cures. In this regard Carpené writes very justly:

"The neutralization of the acetic acid, which has developed in the wine by the oxidation of the alcohol with potash, soda, lime, magnesia, and their simple or double neutral carbonates and tartrates, seems to be a rational method, but, in reality, is not so. These substances neutralize wholly, or in part, the free, and even the combined acids, and the diminution of the complex acidity of the wine renders the acetic taste less noticeable, but does not completely remove it. To remove entirely the acetic acid it is necessary to completely neutralize the wine, because the acetic acid combines with the alkaline and earthy-alkaline bases after they have neutralized the tartaric, malic, and succinic acids. Moreover, acetic acid, even when completely combined with a base, gives out, though less strongly, its characteristic odor, so that even after complete neutralization the wine will still have an odor of acetic acid, accompanied besides by a bitter taste, which lingers in the throat, and may be worse than the first fault."

MILK-SOUR, LACTIC ACID.—This, by inexperienced tasters, is easily confounded with pricking or acetification.

A milk-sour wine has a more disgusting, biting, and penetrating acidity than an acetic wine, a harsh acidity, whose effect is felt long after the wine is swallowed. An acetic wine has a noticeable odor of vinegar, whilst a milk-sour wine emits an odor of rancid butter, due to the butyric acid which almost always accompanies lactic acid.

If there is any doubt as to which acid the wine contains, the doubt can be solved by pouring a drop or two of the wine into the palm of one hand, and then rubbing it with the other; if any acetic acid is present its odor will be immediately perceptible on the hands.

A milk-sour wine loses some of its fluidity, and its color becomes dull.

Sweet, badly defecated wines, especially those rich in albuminoids, are liable to milk-sourness.

The disease appears during the winter or in the spring, and generally in wines poor in acids; it is accompanied by a turbidity of the wine and a change of color. As long as the wine remains in full, well-bunged casks, this turbidity and change of color do not occur, but only when it is exposed to the air.

Some observers have considered lactic acid as one of the normal products of alcoholic fermentation, like glycerine, succinic acid, etc.; the truth, however, is, as Pasteur has proved, that whenever the smallest quantity or trace of lactic acid is found in wine it is caused by lactic fermentation.

Whenever the alcoholic fermentation of certain musts, rich in nitrogenous matters, is not well conducted, especially as regards temperature, a certain quantity of lactic acid is very easily formed, which is a bad defect. This happens generally in certain years in warm countries, where the so-called sweet-sour wines are produced.

It is difficult, not to say impossible, to take away the defect of milk-sourness; the different methods proposed, including that of refermentation, do not succeed; consequently, the best thing is to prevent it by a thorough defecation of the must, and a properly regulated fermentation, not allowing the temperature to rise to a point at which the alcoholic ferment becomes inactive, and thus preventing it from reducing all, or the major part, of the glucose contained in the must.